The Beginner's Guide to Winning the Nobel Prize

The Beginner's Guide to Winning the Nobel Prize.

A LIFE IN SCIENCE

Peter Doherty

Columbia University Press   New York

Columbia University Press
Publishers Since 1893
New York

Library of Congress Cataloging-in-Publication Data

Doherty, P. C. (Peter C.)
   The beginner's guide to winning the Nobel Prize : a life in science / Peter
Doherty.
      p. ; cm.
   Includes bibliographical references and index.
   ISBN 0-231-13896-2 (alk. paper)
   1.   Doherty, P. C. (Peter C.) 2.   Immunologists—United States—
Biography. 3.   Nobel Prizes—Biography. 4.   T-cells—Research—History.
I. Title.   [DNLM: 1.   Doherty, P. C. (Peter C.) 2.   Allergy and
Immunology—Personal Narratives. 3.   Biomedical Research—Personal
Narratives. 4.   Nobel Prize—Personal Narratives.
WZ 100 D655b 2006]
QR180.72.D64A3 2006
616.07′9092—dc22

                                                            2005032326

⊗
Columbia University Press books are printed on permanent and durable acid-
free paper.
Printed in the United States of America
c 10 9 8 7 6 5 4 3 2 1

Contents

Contents

Acknowledgements

This book is intended for a general readership that may not necessarily know much about, or even have sympathy for, the world of science. It would not have existed if Mary Cunnane had not offered her services as a literary agent and, as the true professional she is, made the telling point that although speaking and writing for ephemeral formats is fine, nothing enduring or comprehensive can come of that approach. Even with Mary's patient support, it took Louise Adler of Melbourne University Publishing to finally induce me to make a commitment and embark on the process. Louise also suggested the title, and has provided resolutely good-humoured support and advice throughout.

Though I have written hundreds of thousands of words that have been published in various science formats over the past forty years and have also, since the Nobel Award in 1996, composed articles and commentaries for newspapers and magazines, I quickly discovered that I was a total novice when it came to the business of creating an interesting and readable book.

Two experienced professionals, editorial consultant Kristine Olsson and Melbourne University Publishing's Sybil Nolan, took the 70,000 words or so of my original draft, re-ordered them, consigned whole chunks to the waste basket, then drew me out on various themes and forced me to extract more personal memories and stories from my own memory banks and the files that my wife Penny has organised and kept since 1996. Many of the more personal reminiscences are as much Penny's as mine: we shared the same experience, but remembered different bits. She has also edited and commented on everything that is written here. Michael Doherty critiqued the paragraphs on Parkinson's disease and schizophrenia. Though most of the ideas, the discussion and 99 per cent of the words in this book are mine, I had a great deal of very high quality direction and encouragement.

Most of the material reflects my own perceptions that have been formed both by my years in the scientific community and by my passion for reading broadly, particularly biography and history. Many of the vague recollections and half memories were checked on a variety of websites accessed via Google, and from books on our own shelves like David Marr's excellent biography of Patrick White. My thinking about the future of science was greatly influenced by years of reading "News and Views" summaries in both *Nature* and *Science*, which is about all any scientist can hope to keep up with outside his or her own specialist field. Talking with colleagues has also been a big help. In particular, the stories about the new Asian Pasteur Institutes came from a chance dinner conversation with the Director, the eminent immunologist Philippe Kourilsky. Sherwood Rowland helped clarify some of the confusion in the complex area of global warming. John Burns and Tony Klein provided useful insights into the lifestyles of mathematicians and physicists.

A great deal of the information about the Nobel Prizes and the other laureates is taken directly from the Nobel website (http://nobelprize.org/). This carries the citations, the presentation speeches, the brief biographies and the Nobel lectures of every Laureate since 1901, together with a lot of other supportive material. The website is a function of the Nobel e-museum, the brainchild of Nils Ringertz, the chair of the Nobel medicine committee who called us on that early October morning in 1996. He immediately became a good friend, and we were delighted to see him again at the 100th anniversary celebrations in December 2001. With a sense of great loss, we learned that he died suddenly in 2002, in his 70th year.

Finally, I apologise in advance to other professional scientists and to 'serious' scientific reviewers who may read this book. Science is vast, and the present account is written largely from the viewpoint of an experimentalist in the medical sciences. Even within those limits, some who work on topics that I've mentioned will undoubtedly read what is here and say to themselves: 'well, he got that half right'. This is at a level that is meant to intrigue a general reader. Anyone who wants more information about a particular science story needs to look in much greater depth. I also say little about the challenges in physics, economics or chemistry that are way outside my area of interest and expertise.

Scientific Terms

AIDS
: Acquired Immune Deficiency Syndrome, caused by HIV

ALL
: Acute lymphoblastic leukaemia

Antibodies
: Secreted proteins that mediate specific immune protection

Antigens
: The structures recognised by antibodies

B cell
: B lymphocyte, the precursor of the antibody forming plasma cell

BSE
: Bovine spongiform encephalopathy, or 'mad cow' disease

CD
: A classification system for molecules of interest to immunologists

CD4$^+$ T cell
: The 'helper' T lymphocyte

CD8$^+$ T cell
: The 'killer', or cytotoxic, T lymphocyte

CFC
: Chlorofluorocarbons that cause ozone depletion

CNS
: Central nervous system, i.e., the brain and spinal cord

CSF
: The cerebrospinal fluid that bathes the CNS

CTL
: Cytotoxic T lymphocyte

DNA
: Deoxyribonucleic acid, the stuff that genes are made of

EBV
: Epstein Barr Virus, the cause of infectious mononucleosis and some leukaemias

GMO	Genetically modified organism
H2	The mouse transplantation system
HIV	Human immunodeficiency virus, the cause of AIDS
HLA	The human transplantation system
HPV	Human papilloma virus, causes cervical cancer
Ir	Immune response genes, coding for molecules recognised by CD4$^+$ T cells
LCM(V)	Lymphocytic choriomeningitis (virus)
Lymphokines	Secreted proteins that promote immunity
mAb	Monoclonal antibody
MRI	Magnetic resonance imaging
mRNA	Messenger RNA that carries the information from the genes
MS	Multiple sclerosis
N	Influenza neuraminidase protein
RBC	Red blood cell
RNA	Ribonucleic acid
T cell	Thymus-derived lymphocyte, or T lymphocyte
T lymphocyte	White blood cell that differentiates in the thymus, same as T cell
TB	Tuberculosis
TCR	T cell receptor
Th1 and Th2	CD4$^+$ T cells with different functional characteristics

Preface

The aim of this book is to give a sense of the world of science from both inside and out. Though this account isn't meant to be an autobiography, I've used episodes from my own life to probe the extraordinary story of Nobel-level science and what shapes and feeds it. This is about a certain type of life that is concerned with asking questions and seeking answers that hold up on further testing and scrutiny. Can we approach the truth, even if it is only a very small truth? Of course, research scientists aren't the only people who question and look for truth and understanding. Many different and extraordinary people are dedicated to illuminating basic truths and promoting the causes that link, rather than divide, us as human beings. Few of them win a Nobel, but these prizes and all they symbolise provide a useful means of focusing on this type of achievement.

Science cultures differ in many details, but the underlying focus is always on discovery and innovation. The very basis of science is about probing reality. Alfred Nobel was an industrialist, but he was deeply involved in discovery and innovation. His great experiment was to provide reward and recognition for major scientific and humanitarian achievements, to celebrate knowledge, compassion and insight. His aim, if you like, was to infect the world with the need for understanding, with a passion for real and creative solutions.

The extraordinary experiment of the Nobel prizes has now been going for more than a hundred years. Nearly ten years after receiving the award, I set out here to look at the lessons from Nobel's experiment. What factors lead some of the best and brightest in a society to commit to lives driven by Nobel's aims—the linked cultures of creativity, knowledge and humanitarian activity? What is at stake for human society in the way research science is now practised, and in the way the dollars and opportunities are allocated and distributed? What about exciting new fields like nano-technology and genomics? Where has science and tech-nology brought us, how did it develop, and where will science take humanity through the twenty-first century? Where does science fit in the history of us, and how does it relate to other great human themes like religion?

Who are the scientists, and what do they have in common? What forms these people, how do they work and what types of lives do they live? Is science a long-term option for someone who wants to have a job, earn some money and raise a family? The truth of it is that most sci-entists do marry, have kids, live in cities and go to work each day much like the rest of the community. Even so, this is one of those jobs that demands commitment and doesn't necessarily make for a 'relaxed and comfortable' way of life.

Nobody can simply decide to win a Nobel Prize, of course, and the chance that buying this book will lead to a Nobel is as remote as the possibility that reading Jack Nicklaus's *Golf My Way* will result in winning the US Open. Such things can happen but, just as any player can enjoy golf, there are other rewards. The decision to work hard and commit to a life based in rational enquiry can bring genuine excitement and a sense of real achievement, if not global accolades, to many. Life at its best is an adventure, a voyage of discovery. What could be more gratifying than to discover, describe and explain some basic principle that no human being has ever understood before? This is the stuff of true science. Those societies that foster and harness that passion will be the prosperous, knowledge-based economies of the future. Most of us cannot, or would not, wish to be scientists. Can we afford, though, to be in ignorance of how science works and what it can achieve?

Preface to the American Edition

ecause the advances that result from science and technology are so profound and have such enormous effects on how we work and live, it is important for each of us to have some understanding of this vast and dynamic enterprise. Why would any young person decide to be a scientist, and why should you be happy if your son or daughter chooses to be one? It helps to know something about where scientists come from, how they train, what they actually do, and what kind of people they are. Is this a good and substantial life, or are scientists the mad, bad, or naive nerds that we too often encounter in Hollywood movies? Are young scientists taking a vow of poverty? Where is science going, and what might someone who enters the profession now be doing in twenty or thirty years?

Science is about the natural world, and, though it is infinitely fascinating to the engaged practitioner, accounts that deal only with the cold, hard facts of evidence-based reality can sometimes be intimidating, and even scary, to those who aren't accustomed to thinking in terms of ideas, experiments, data, and critical scrutiny. The Nobel Prizes, though, are also about people, so relating a little of my own experience and the lives of others who have been recognized in this way does, I hope, put a human face on what is, after all, among the greatest of all human adventures.

As we look back on the twentieth century, a source of great pride for every American should be the enormous advances in human well-being that have resulted from the massive research enterprise supported by the U.S. National Institutes of Health (the NIH). This federally funded organization operates in two parts, the intramural effort at the vast complex of research laboratories on the main campus at Bethesda, Maryland, and the extramural program funding the investigator-initiated, peer-reviewed grants that support health research in universities, hospitals, and dedicated institutes throughout the United States. If you want a single reason why the majority of the Nobel Prizes in Physiology or Medicine over the past forty years have gone to scientists working in the United States, think NIH.

The generous support given to the NIH by the U.S. taxpayer is what brought me, and many other Nobel laureates, from the countries where we were born and grew up to head research laboratories here. In 1975 I left the Australian National University in Canberra to join Philadelphia's Wistar Institute, embedded in the grounds and the intellectual ambience of the University of Pennyslvania. On one side of Spruce Street is America's oldest, private biomedical research institute; on the other, Ben Franklin's original university medical school. Both are great institutions.

After seven years I went back "down under" for a time but returned in 1988 to become head of the Department of Immunology at St. Jude Children's Research Hospital in Memphis, Tennessee. I still hold the Michael F. Tamer Chair of Biomedical Research at St. Jude, though I split my time between Memphis and Australia's University of Melbourne. Until the 2005 Nobel medicine prize was given to the Perth-based researchers Barry Marshall and Robin Warren, I was the only living Nobel laureate in the sciences spending even part of the year in the Southern hemisphere. Such is the north/south divide in both wealth and intellectual achievement.

My wife, Penny, and I were living happily in Memphis when, very early on an October morning in 1996, I got the call telling me that I was to share the 1996 Nobel medicine prize with my Swiss colleague, Rolf Zinkernagel. This propelled me into a new and different world of public advocacy for science. St. Jude is a full-time biomedical research operation focused on finding solutions to catastrophic diseases in children. Many Americans are, of course, aware of this marvelous institution because they have either donated money or been involved in fund-raising events like trike-a-thons and math-a-thons. Most will be familiar with the TV programs about St. Jude patients organized by Marlo Thomas, the elder daughter of the hospital's founder—the actor, producer, and humanitarian Danny Thomas.

After winning the Nobel, I immediately became part of the St. Jude publicity machine and found myself doing double bills with the glamorous Marlo at various venues around the country. Marlo is, of course, a well-known actress, who speaks with skill and sincere emotional weight. I can never remember what I've said from one moment to the next so had to make up a fresh story every time I spoke. We were, I think, mutually impressed by the very different attributes we brought to the process of seeking better outcomes for desperately sick kids. Marlo's brother Tony produced two of Robin Williams's movies, including *Dead Poet's Society* and *Insomnia*. We met Robin when he performed a comedy act about my Nobel at a Hollywood fund-raising event organized by Marlo and asked him about *Mrs. Doubtfire*. Years before, in another life, Penny and I bought a bed from a secondhand store in Edinburgh run by Madame Doubtfire, who, as Robin confirmed, provided the name for the film.

We went to Monaco and socialized with some of the glitterati on their home turf. (Despite a ban on photographs, Marlo's husband, Phil Donohue, used the excuse of demonstrating a new digital camera to sneak a picture of Penny with the risqué and charming Prince Albert.)

Memphis also hosts the annual meeting of Country Cares, an organization of performers and radio disc jockeys from all over the United States that raises millions of dollars each year for St. Jude. As a backwoods Australian of Anglo-Irish descent, I had no trouble convincing these infinitely good-hearted, open, and generous people that I was just another down-home "good ole boy." This was a year like none I had experienced.

Most U.S. laureates have a Nobel year, which takes them away from their research and puts them on the public stage. When the next awards are announced, the pressure on them tends to fall off, and, if they aren't too old, they generally return to work within two years. It proved to be more complicated for me. Apart from being the first St. Jude scientist to win a Nobel, I was also the only Australian since 1975 to be recognized by any of the Nobel committees. Much to my surprise, I found myself named the 1997 Australian of the Year, a distinction that led to numerous trips across the Pacific to speak in many different venues. I discovered that I had something of a flair for communicating the excitement and value of science to a broad audience, a talent that was sorely lacking on the Australian scene.

The problem with scientists engaged in public advocacy is that, unlike the reponse they receive from a research paper or a review article, they have little sense of where the message has gone, if it is effective, or even if anyone has heard it. Scientists are used to dealing with hard data, but the communication business gives one the sense of being involved in a transient, ephemeral activity. What induced me to write *The Beginner's Guide to Winning the Nobel Prize* was a very personal need to bring together what I had been pushing on the public stage in an accessible and enduring format. This is my first book about science, but I hope it will not be the last.

Being a part-time speaker can be both gratifying and exhausting. Writing the book caused me to think in greater depth about what I had been saying and what I was actually trying to achieve as a communicator. In the book I've deliberately used a fairly broad brush and—apart from talking about the basic principles and history of contemporary science—have ranged from my own, specific areas of research in immunity and influenza to topics such as plant genetic engineering, global warming, renewable energy, cancer, genomics, behavioral neuroscience, and medical ethics. Given the attention paid to the subject for the last few years, I've also included a chapter on science and religion.

This little book was fun to write. I hope you enjoy reading it.

The Beginner's Guide to Winning the Nobel Prize

Introduction

We were living in Memphis, Tennessee, when the phone rang at 4.20 one cool October morning. My wife Penny picked it up, thinking there could be a problem with an elderly parent back in Australia. But the voice wasn't Australian. 'This is Nils Ringertz', she heard, 'from the Nobel Foundation'. Penny handed me the phone. 'This is for you', she said.

Down the line from Sweden, Nils told me that I was to share the 1996 Nobel Prize for Physiology or Medicine with my Swiss friend and colleague Rolf Zinkernagel, for a discovery we made more than twenty years previously. He also warned us we had ten minutes to call our family before he made the announcement to the press. The phone, he said with some understatement, would then be constantly busy. As I recall, we were in mild shock.

I had known for some time that I might be in the running for the Nobel, but those sorts of rumours had been circulating for years and I hadn't taken much notice of them until very recently. A year previously, Rolf and I had shared in the Lasker Basic science award, a prestigious American prize that tends to predict future Nobels. Some of my racier colleagues actually had me at 30 per cent odds for a trip to Stockholm, but I wasn't too excited. As much from the viewpoint of psychological self-preservation as anything else, I had myself convinced that boys from the Australian backblocks don't win Nobel Prizes. That morning, though, there was no doubt. Within fifteen minutes we were fielding calls from Reuters, Belgium, talk-back radio in Bogota, Colombia, the *Sydney Morning Herald* and so forth. Our telephone records show that we got one call out at 4.27 am and the next we managed was at 5.32 am. This clearly wasn't going to be an average Monday. Life hasn't, in fact, been quite normal since.

Of course, everyone's idea of 'normal' is different. As a child growing up in the sub-tropical city of Brisbane, I may not have believed a life of science—spent mainly in laboratories, between three continents—was normal, either. Childhood in Queensland in the middle of the twentieth century was a fairly quiet and unintellectual affair. I had little idea of what the wider world was like and not a whole lot of information to go on. Looking back, it hardly seems the kind of springboard that would catapult anyone into the higher echelons of discovery.

I grew up in the outer working-class suburb of Oxley, where half the students at my local primary school left at the end of eighth grade to work in the local 'bacon factory'—a pig slaughterhouse—the cement factory, the brick works, or to take up apprenticeships. Though I was a bright kid, my school days moved slowly; I was often bored and under-performed. It didn't help that I was weedy, poorly co-ordinated, and a year younger than almost everyone else. I tried, but I was a liability for the side in any competitive sport.

Things improved a lot when, at age 13, I got to high school. It was a brand new facility that opened the year I entered, so there were no older students to provide an example, no library to speak of and no student clubs. What saved me were university-educated teachers who were totally dedicated to the idea of public education. Streamed into an academic class, I got a good grounding in physics, chemistry and maths, and a love of history and the great books and plays of the English language. My first introduction to a foreign culture was high school French. Though my spoken French is terrible and I no longer read it too well, the exposure to French history and culture was an eye-opener. I take pride in the fact that, after the Nobel, I was elected as foreign associate of the French Academy of Medicine.

Back then, Brisbane was a rather isolated and parochial town in a country barely noticed by the rest of the world. My views as a youngster were formed by reading—though the only reference to the United States in my history book was a short chapter entitled 'George III and the loss of the American colonies'—and movies. I ended up with a view of US history that was both Anglicised and influenced by John Wayne. That barely changed when, in 1956, the year before I started university, the first television transmissions began in Australia—more westerns, with Australian game shows thrown in. Television provided no more illumination about our nearer neighbours, either: the little we learned about the Asian countries to our north related to World War II and the European colonial experience.

Nor did my family background provide many hints about what was ahead for me, or what path I might follow. My parents had both left school at age 15 though, like many in their generation who had limited formal schooling, they spoke clear grammatical English and could write a lucid letter. My mother had continued with lessons to become a piano teacher, and the house echoed to Debussy, Chopin and Mozart. My father took a variety of 'in service' courses in his job, initially as a telephone technician and later on the management side of telephone services. He was an avid reader of anything and everything. However, they had no understanding of higher education. In fact, very few people in the area had a university degree, except for the local doctor and dentist; there weren't too many obvious people to turn to for career advice. Oxley, with its weatherboard houses on stilts and a semi-rural feel, was one of Brisbane's peripheral 'struggle towns'.

I had two friends in an adjacent, more prosperous suburb whose fathers were in professional life, but it never occurred to me that I could discuss education and careers with them. Then there was my cousin, Ralph Doherty, who was thirteen years older and lived on the other side of the city's massive sprawl, was very bright and topped the state academically. He was the first in the extended family to go to university and he graduated with great distinction from the University of Queensland Medical School, eventually going into tropical public health and infectious disease research, and then on to Harvard for post-graduate study. I was vaguely aware of this, but don't recall ever having a serious conversation with him about science. Besides, it was assumed that Ralph was so super-smart that nobody could hope to emulate his example.

After high school, I had no clear idea of what I might try, though one possibility I did consider was becoming a cadet journalist on the local newspaper, the Brisbane *Courier-Mail*. I was reading avidly. Reading the French existentialist philosopher Jean-Paul Sartre introduced me to the age of reason. At the same time, Aldous Huxley's novels, such as *Eyeless in Gaza* and *Point Counterpoint*, that interweave some of the scientific themes of his day (the 1920s and 30s) with the lives of his upper-class English characters also brought me into contact with a culture that looked to the Enlightenment and the evidence-based world of scientific research. Huxley used current thinking in developmental biology, for instance, to develop storylines that explore the tension between passion and the life of the mind. What normal 16-year-old is not interested in passion? I hadn't studied biology at school—it wasn't offered to boys for, I suspect, much the same reason that some religious conservatives now object to sex education—but the idea of doing research in some field of biology looked interesting. How should I go about this? I didn't want to train as a medical doctor because, so far as I knew, most of them spent their lives dealing with sick or neurotic people. This didn't sound like much fun to me.

What changed my life was going to an 'open day' at the University of Queensland School of Veterinary Science. At that time, the 'U of Q' was one of only two places where veterinarians could train in Australia and New Zealand. My interest was immediately piqued by the demonstrations in embryology, anatomy and pathology, and by the rather scatty, sexy, chain-smoking young laboratory technician who looked after the displays. In the hot Brisbane summer, she wore a white laboratory gown and not much else. This 'older woman'—she must have been all of twenty-two— certainly wasn't the self-important Dr Frankenstein of the movies in the carefully buttoned white coat. Even the diseased organs displayed around the walls and the permeating smell of hot embedding wax and formalin were intriguing. This was all so different from anything that I had ever encountered in my sixteen years. It looked real and, above all, interesting and doable. From that moment I was hooked on pathology.

Pathology is clearly a turn-on for adolescents. Many young people elect to study forensics after watching those gruesome television programs with 'floaters', electric bone saws and hard-nosed characters who spend a good part of their lives in white plastic overalls, snipping off bits and putting them in bottles. I retain my fascination with disease and death even now: yes, it's true that many innovative research scientists are stuck in a state of perpetual adolescence. The 'disease detective' game constantly turns up surprises, and it certainly isn't boring.

Medicine, dentistry and veterinary medicine are postgraduate courses in the United States but, at least in those distant days, Australia, like Britain, started all young people into professional training straight out of high school. If I had gone first to a US four-year liberal arts college, I would probably now be both better educated and an historian. Even as a scientist, I always tend to develop explanations by beginning from a historical perspective and am fascinated by history and politics.

I began at the veterinary school at age 17 and graduated five years later in the bright, hot summer of December 1962. Exactly thirty-four years later, in December 1996, I found myself in bleak, wintry Stockholm receiving the Nobel Prize for Physiology or Medicine from the hand of King Carl XVI Gustaf of Sweden. What took me from a young, naïve vet student to immunology and the kind of discovery science that occasionally turns up results that win prizes? There weren't that many differences between my fellow students and me back then, but one was that I always wanted, from the outset, to be a research scientist. I was altruistic enough to believe that improving the health of domestic animals, so important in the developing world, would be something worthwhile. After my graduation, rather than go into veterinary practice, I worked on infectious disease problems in cattle, pigs, chickens and sheep, first in Queensland and then in Scotland, where I completed my PhD on louping ill encephalitis, a tick-borne virus-induced brain inflammation in sheep.

My long-term aim after Edinburgh was to be a veterinary researcher with the large, national, applied research organisation, the CSIRO in Melbourne. First though, I diverted—I thought temporarily—to the John Curtin School of Medical Research (JCSMR) at the Australian National University to learn about cell-mediated immunity, so that I could better understand the host response to viruses. In Canberra, I started my experiments on virus infections in laboratory mice in 1972 and was introduced, for the first time, to a dynamic, intellectually driven, basic medical research environment. The story of what happened next in my science odyssey is told later in this book. Needless to say, I never made it back to work in the veterinary world.

I have since worked in both Australia and the United States, but won the Nobel Prize for a discovery made in Canberra, and for the intellectual framework Rolf Zinkernagel and I developed there during 1973–75 to explain our findings. Within a couple of years, we were both being recognised as significant figures in the world of immunology, a status that we have maintained. The Nobel, of course, moves that business of fame and reputation into a different league. The initial, intense global media attention doesn't continue much beyond the award week in Sweden, but the recognition, I now realise, lasts much longer and permeates the rest of your life. 'Nobel Prize winner' is a permanent job description. The continued reputation rests, of course, as much in the status of the award as in the achievements of an individual Nobel laureate.

What would the naïve, unsophisticated Oxley schoolboy have thought if he could have peered into a crystal ball and observed himself in Stockholm those years later, looking out from the Grand Hotel to the Royal Palace? What if someone had told him that an international career and one of the world's most prestigious prizes awaited him down the line? I'm not sure I was even aware of the status of the Nobel then, or the names of my countrymen who had received it. Winning a Nobel wasn't what I set out to do with my life, and as far as I was concerned, it was an extraordinarily improbable outcome. Why me?

My personal view of the prize is that, like many science laureates, I was recognised for my part in making a breakthrough discovery that changed the prevailing view, what the philosopher Thomas Kuhn calls a 'paradigm shift'. We did some rather simple experiments and advanced what was at that time a revolutionary explanation for our results. Many outstanding scientists then used technological advances in other fields to explain both what we had found and what came after. Their stories are no doubt just as interesting as mine, and every one will have been influenced by many factors, including people, places, opportunities and intellectual environments. Though few win Nobel Prizes, all who work at the forefront of discovery and problem-solving are part of that same tradition, whether they be scientists, writers or peace-makers.

1

The Swedish Effect

Stockholm in December: darkness falls early like a frozen curtain, the short days are dimmed by snow-fall, and even weather-hardened Swedes grimace in the winds that cut across Strommen, the waterway straddled by the city. I grew up in a hot, humid, place where the sun shone pitilessly and I was always getting burnt. As a consequence, I'm energised by cold, bleak, misty weather. Perhaps it's a heritage that goes back to ancestors who dug in dank Irish peat bogs, herded hill sheep in a Lancashire winter, or lived by a canal in Essex, but cold damp Stockholm in December suits me just fine. There is a freshness to the air, a sense of possibility.

Despite their impatience with the long winter, 10 December—Nobel Prize Day—is an exciting occasion for Swedes young and old. Like Australians on Melbourne Cup Day or Americans on Super Bowl Sunday, Swedes will stop whatever they are doing to watch television, or try to catch a glimpse of the proceedings at the Stockholm Concert Hall and the Town Hall—and it isn't just the spectacle of the king and queen and hundreds of people in formal dress. Swedes are immensely proud of their culture and traditions, and there is a great identification among them with a Scandinavian tradition of democracy, consensus and fairness. So the ceremonies surrounding the annual Nobel Prize presentations are for most an important reminder of great human endeavours, and of their own nation's place in promoting them. The awards have a high profile throughout the country: Swedish television carries live telecasts of the presentations and banquet, newspapers and radio programs feature the winners, and people talk about the prizes on the street.

Until they win, few laureates realise that the award ceremony is associated with an intense but exhausting week that thrusts them suddenly into the media spotlight, and requires a high level of energy and—unless they are teetotal—a reasonably tolerant liver. Neither would they anticipate the other accoutrements, including a chauffeur and a junior Swedish diplomat assigned as helper and advisor. Scientists, at least, don't normally live in a world of minders and personal limousines—certainly not one of celebrities. However, when the king confers that award, handing the winner a gold medal and a leather-bound certificate in an atmosphere of solemn dignity, he also bestows a kind of celebrity status that has its own rewards and limitations.

The latter mainly involves the loss of personal and professional time that goes with public attention, but the compensations are the broader awareness of your work, gaining a public 'voice' and the opportunities to meet extraordinary people. I had a lot of media coverage in Memphis at the time of the award, and am on radio and television from time to time in Australia. A few people recognise me in the street. So far as popular celebrity goes, I've likened it to being equivalent to that for someone in a crowd scene in a television coffee commercial. It may be different for those few Nobel laureates who are recognised while they're still young and handsome, but most of us tend to look more like the frog than the prince.

The presentation ceremony at the Stockholm Concert Hall is followed by the white-tie Nobel banquet—complete with gold-leafed plates and gold-plated cutlery—for 1200 in the town hall. Protocol is paramount: laureates are taught how to bow, and are told it is all right to turn their back on the king as they return to their seat (we were shown a video of a statuesque Pearl Buck, the 1938 literature winner, reversing very uncomfortably in a tight evening dress for what seems about a hundred feet). Laureates are also taught how to make the Scandinavian toast 'skal', and the men must wear full evening dress for the ceremony and dinners.

At the banquet wine is drunk from Orrefors crystal glasses, of course. The crash when large numbers of plates hit the table simultaneously is deafening. Waiters volunteer from all over Sweden and are thick on the ground, but as the Nobel Foundation is the host, the comperes and ushers for the evening are young, attractive university students. Places are set aside for both local and international students. There is classical music, a few songs from opera and one representative from each of the prize categories gives a short speech. Rolf Zinkernagel spoke with eloquence and charm for the two of us.

I may have been the only person at the dinner who didn't have the famous Nobel ice cream. Not long before, I'd been diagnosed as having problems relating to the high blood cholesterol that shortened the lives of both my father and grandfather. We have a picture of me with a very elaborate pineapple concoction. Since then, as a consequence of the work that won Joe Goldstein and Mike Brown the 1985 Medicine Prize, I have been taking high doses of the very effective cholesterol-lowering statins. (It seems to be working: when all the laureates who were in good enough health were called back for a repeat dinner at the 100th anniversary of the Nobel Prizes in 2001, I was finally allowed to taste the ice cream.)

After dinner there were photographs, dancing and an audience with the king and queen. The king has the royal prerogative of initiating any conversation and, though we had flown to Stockholm from Memphis, we were correctly identified as being Australian. As I recall, we spoke a little about the southern sky and their recent visit to the Paranal Observatory, which is part of the European Southern Observatory complex in Chile's Atacarma desert. Unlike the Queen of England, the Swedish monarch no longer has any statutory role and his function is essentially symbolic and ceremonial. The royal couple were amiable, interested and interesting.

The following night, the king and queen hosted a spectacular dinner at the Royal Palace. This was only the second time I'd ever worn white tie and tails. Scientists live by Murphy's Law ('Anything that can go wrong will go wrong'), and I discovered that this extends to minor sartorial disasters. Just as we were stepping out of our stretch Volvo at the palace, the elastic holding the starched front that goes under the dark jacket snapped and it shot out like a white, horizontal flag. Perhaps for the first time in her life Penny, who was carrying a very small evening bag, did not have a safety pin. Our driver, Gretel Lundstrom, immediately recognising the emergency, sped us away from the palace grounds in a dramatic exit that apparently caused some alarm for the organisers. We made the short trip back to the Grand Hotel, rushed to our room and repaired the damage with safety pins on the way down in the elevator. Then back to the palace with a police escort and flashing blue lights, followed by a quick dash up the stairs to be the last arrivals.

We crept in to our correct place in the reception line: protocol requires that the medicine laureates and their spouses are behind the chemists, but preceding the literature laureate. Our escapade was an amusing diversion for the rather shy Polish poet Wislawa Zymborska. On the other hand, it taught me that members of the Nobel Foundation, like the Swedish people, are quite unflappable. Over the course of 100 years they have seen just about everything—including at least two laureates who turned up with three wives, past, present and future. Only in the fictitious plot of the 1963 Hollywood movie, *The Prize*, did they ever have to deal with a murder.

The setting for the royal dinner was again formal, though the atmosphere was a little more relaxed than at the awards. The king described it as 'a family dinner'. The servers were dressed as footmen, the one standing behind the queen sporting an enormous feather projecting far above his head. This tradition was evidently started by an earlier king so that he could immediately find the queen in a crowded reception.

My dinner companions were the Swedish foreign minister, Lena Hjelm-Wallen, on one side and Princess Lillian of Sweden on the other. Princess Lillian, the wife of the king's popular uncle, Prince Bertil, was born in Swansea, Wales, and is greatly loved in Sweden. She has a wicked and irreverent sense of humour and, at the age of 81, was a very enthusiastic 'skaaler', if there is such a word. We ate venison that had been shot on the royal estate by the king, and it was in every way an enjoyable evening. Talking with the Swedish heir apparent, Princess Victoria, during her recent visit to Australia, I was delighted to hear that Princess Lillian is still in fine fettle.

During the course of the Nobel week I delivered the lecture that every laureate is required to give, and in the evenings we were entertained by Swedish colleagues in the medical sciences. We had dinner with the Nobel medicine committee members, and spent another evening at the house of the immunologist Hans Wigzell, an old friend who is head of the Karolinska Institute, Sweden's premier medical university which was founded in Stockholm in 1810.

We also found ourselves caught up in the celebrations for Santa Lucia Day, 13 December. This is the traditional beginning of the Swedish Christmas season and a moving example of how close the Swedes hold their customs and festivals. Just how or where the symbolic figure, Santa Lucia, came from is not completely clear. These days no one argues about origins, but in just about every house, office, school and club across Sweden, the Lucia—wearing a wreath of candles in her hair—leads a procession of white-clad girls and boys as they bring light and food and sing their Lucia carols in the early morning darkness. There were even fully lit Lucias, accompanied by choirs singing seasonal songs, at the airport as we headed off in the early morning to Lund on the south-west coast, where I was to repeat the Nobel lecture that I'd already presented at the Karolinska Medical School and the University of Uppsala, north of Stockholm.

Santa Lucia Day ends for the Medicine laureates, their families and friends at a final white-tie dinner with faculty and students from the Karolinska Medical School. Toasts are made, more 'punch' is drunk, there are speeches and Swedish drinking songs from a book printed in English. The Swedes don't drink and drive, so everyone comes by cab or hire car and the punch is fairly lethal. The laureates are expected to cross hands and make a chair to carry in a Lucia girl selected from the first-year medical students.

The students had prepared an entertainment based on the assumption that an American would win but, faced with a Swiss and an Australian, went ahead with it anyway. In turn, with the Zinkernagels, the Australian ambassador, an Australian science journalist, some old friends and a few Karolinska faculty members who had ties to Australia, Penny and I sang a rather loud, but slightly woozy, version of *Waltzing Matilda*. At that stage of the evening, we all benefited from the fact that a complete version of it was in the songbook. The Nobel preparations cater for every possibility.

As new—and in some ways overwhelming—as it is, the experience of the Nobel week stands you in good stead for the reality of the very busy 'Nobel year' that all laureates face when the celebrations in Scandinavia end and they return home. You can avoid the endless rounds of engagements, of course, by either refusing invitations or by giving talks so appallingly bad that even the Nobel cachet cannot improve them—the word gets around. Most people take the responsibility seriously because it provides the opportunity to air issues they care about in a broad, public forum. The pressure generally tends to tail off after the next series of awards is announced, though there are always invitations that seem to relate purely to the status of the

Nobel Prize. The Nobel year's commitments consume a lot of time and take active scientists away from their research programs. Some, who are in the later stages of their careers, lose traction and never really get back to what had been their life's work. All laureates, at any rate, lose a measure of the personal space required for introspection and creativity.

My year, however, like that of most senior scientists, was essentially booked and committed before the call came from Stockholm. Apart from my own ongoing work at St Jude Children's Research Hospital in Memphis, I was serving on various national committees and was booked to give numerous seminars. The Nobel adds a further level of invitations, many of which cannot be refused. I received Australia's highest civil award, the Companion of the Order of Australia, from the then Governor-General, Sir William Deane, and gave masses of lectures associated with visits 'back home', including an address to the National Press Club in Canberra that was broadcast over and over. I was constantly giving public talks and appearing at public events while in Australia. There was never a spare half-day, or so it seemed. The publicity also draws old friends back into contact, while the various institutions and groups that can claim some affiliation take pride in that and issue invitations accordingly. In order to cope, I chose to drop out of the time-consuming work of reviewing grants and manuscripts, and put off planned seminars at several universities. I also had to excuse myself from the review committee for the US Multiple Sclerosis Society, which was close to my heart. They had funded me in the past, and a dear friend had committed suicide years before because of her MS. It was all pretty exhausting, and even now I still have to be extremely judicious about the commitments I accept.

I am one of only two Australian Nobel laureates still living, and these days spend much of the year in Melbourne. Almost ten years on from the award, I continue to receive invitations to speak and appear at numerous functions. Australian laureates have been fairly scarce on the ground over the past 100 years: just seven grew up and at least completed high school here before leaving for places north.

Just about everyone has heard of the Nobel Prizes but, unless they are Scandinavian, most don't know much about them. Deciding who should receive a Nobel is a considered and protracted process and is, even now, informed by the life and intentions of Alfred Nobel. The life of this wealthy industrialist and explosives manufacturer presents intriguing paradoxes. He invented the detonator, dynamite and a smokeless powder to propel cannon shells—all agents of destruction, in some sense—but many believe he intended all the prizes to operate to promote world peace. In that sense, they are all 'peace prizes' and clearly the words 'peace' and 'Nobel' go together in the broader public consciousness. This is driven home to me when I am occasionally introduced as a recipient of a Nobel Peace Prize. It is likely that Nobel the man—who never married and had no children—would now be largely forgotten if it were not for the extraordinary legacy he left, and the precision with which it has been administered.

Before he died on 10 December 1896, Nobel instructed that the majority of his estate should be invested so that the accumulating interest could be used to fund what we all know as the Nobel Prizes. In his will he also specified the disciplines to be recognised, the institutions that should be responsible for decision-making, and guidelines about who should win. The most important single factor determining the status of the Prizes is the statement in the last sentence of the extract from Nobel's will that is published for all to read on the Nobel website. Translated from the Swedish, it reads: 'It is my express wish that in awarding the prizes no consideration be given to the nationality of the candidates, but that the most worthy shall receive the prize, whether he be Scandinavian or not'. Here we have it. These are international awards that try to identify the most important, the most significant work in physics, chemistry, medicine or physiology, literature, economics and peace, by individuals who have contributed most substantially to benefit humanity.

Nobel's executors were two young engineers who worked for his company—Ragnar Sohlman and Rudolf Lilljequist. Sohlman was instrumental in establishing the Nobel Foundation that is responsible for both the investments and the awards. The intent and scope of the prizes generally follows the brief instructions in Nobel's will, but practical considerations have required further interpretation by the Nobel Foundation. One consequence is that the award is given to a maximum of three people in each category, with the exception of the Peace Prize that can go to large organisations like the Red Cross or Médecins sans Frontières.

The Literature Prize is decided by a committee drawn from the Swedish Academy, founded in 1786; the Physics and Chemistry Prizes by separate committees from within the Royal Swedish Academy of Sciences, founded in 1739; the Prize in Physiology or Medicine by members of the Karolinska Institute, which boasts its own alumni of Nobel winners; the Peace Prize by a committee selected by the Norwegian Parliament, the Storting. Although there is no prize in mathematics, the Nobel Foundation added the Bank of Sweden Prize in Economic Sciences in memory of Alfred Nobel in 1969. Responsibility for the Economics Prize—which some still consider to be controversial—was handed to the Royal Swedish Academy of Sciences, which sometimes chooses a mathematical economist.

Invitations to potential nominators go out in the later part of the year and they must forward any nominations they want to submit by the following February. The list of nominees is potentially enormous and varied, so there are strict rules around those eligible to nominate. For the science prizes, the general protocol is that major institutions and individuals prominent in particular areas are asked to provide names, a concise justification and background information. Though former laureates are always invited to submit nominations, they are restricted to the field for which they were recognised. I enquired during Nobel week if I could propose a name in another area, perhaps literature. The answer was a definite no. One experience that Nobel laureates share is the occasional letter from someone who is, to be polite, 'way out in left field' and believes he (it's always a he) merits a Nobel Prize and should be nominated. I've kept these for posterity.

By October, the committees make their recommendations, and their decisions must be ratified by the broader body of the responsible academy. The discussions and deliberations involved in this painstaking process remain confidential, and are embargoed for fifty years. Whenever I read that someone has had the distinction of being 'nominated for a Nobel Prize', I wonder how the information emerged. Candidates in the sciences are most unlikely to be aware that they are being considered. They might be asked to provide an up-to-date curriculum vitae, or a list of what they consider to be their achievements, but the enquiry is likely to have come from a third-party source that would not normally spark the thought of a Nobel connection. The nominators are asked to be as circumspect as the committees that make the decisions. Though the scientific community gossips about possible future winners, and everyone assumes that the obvious people have been nominated, the details are generally known only to the proposers themselves.

The entire process is secretive, and the Swedes never leak. On the other hand, you do hear rumours that you've been nominated for a Nobel. I'd never tried to pursue this in any way—perhaps because, years ago, I was told by organisers I'd been nominated for another prestigious prize and became a bit fixated on it. Then nothing happened, so I've ignored anything like that since then.

Some people get very distressed about the Nobel because they think they should win. However, there is no 'court of appeal' and some never really recover from feeling they've been excluded. There is also the 'rule of three', which means that tough decisions must be made. At least in the medical area, people will have been nominated a number of times for different awards. Anyone who has a case will have been looked at pretty closely. I suspect that the same Nobel candidates come up year after year and, as I understand it, the leading cases are reviewed afresh each time.

Half the people who, like us, win a Lasker Basic science award later win a Nobel. American culture is very open, and being American, the Lasker committee leaks like a sieve. As I heard it, Rolf and I were added to a 'Harvard ticket' of Don Wiley, Jack Strominger and Emil Unanue after the immunologist on the committee said it would be outrageous to recognise them and omit us. Don and Jack provided the X-ray crystallography pictures that explained how our 'altered self' idea worked (see Appendix 1), while Emil worked on a related problem with 'helper' T cells (which is discussed in chapter 4).

While the prize itself affects each individual differently, it's worth thinking about the general effect of these awards on Sweden's place in the world. In the end analysis, the biggest, long-term winner from the Nobel Prizes may be the land of Alfred Nobel's birth.

To my mind, the Nobel Prizes have proven to be a great device for focusing the broad attention of society on values that relate to rational, evidence-based enquiry, truth and peace, the basic building blocks of prosperity and of participatory democracy. Sweden is a modern, industrialised nation. The country spends a very high percentage of gross domestic product—more than 4 per cent in 2001—on research and development. This combines commitments from the public and the private sector. And it shows: anyone who has worked in medicine over the past fifty years is likely to have used products from the two large Swedish pharmaceutical companies, Astra and Pharmacia. By 2002, little Sweden had the fourth largest biotechnology industry in Europe. All of us are familiar with names like Volvo, Scania and Saab when we think of cars, trucks and fighter planes.

My father, who started his working life as a telephone mechanic, used equipment made by Ericsson, the international giant based in Stockholm that continues to drive advances in telecommunications, broadband and the like throughout the world. The engineering industry accounted for 9 per cent of industrial output in 1900, 23 per cent in 1945 and 50 per cent in 1999. Sweden has also had a major armaments industry, which might be thought of as another legacy of Nobel. The Bofors anti-aircraft gun, for example, was used by the military on both sides in World War II. Art and design don't miss out, either: witness the huge success of the Swedish glass industry and the Ikea homewares juggernaut.

It makes sense that the country that produced Nobel pays a lot of attention to its scientists and their institutions. Others do too: because the selection committees for the science prizes consist solely of people working within the responsible academies, ambitious people who believe they may have some chance of a Nobel will invite their Swedish colleagues to conferences and gatherings where their work is on show. In turn, the Swedish scientific community is very conscious of this and the standard of biomedical research at the Karolinska Institute, for instance, is extremely high. The population of Sweden is about 9 million, and five scientists from the Karolinska have been awarded the Nobel Medicine Prize. A surprising number of my friends doing medical research, in both the United States and Australia, have spent time working at the Karolinska in their younger days, simply because it's a great place to do science.

Though the modernisation of Sweden is generally considered to have begun in the 1870s, a move forward that included the personal efforts of Alfred Nobel, it is clear that the past 100 years have seen a major change in the country. Rapid technological advancement has been accompanied by reasonably good relations between employers and trades unions. The university system is first-rate. It would be hard to measure, but it seems likely that the strong political commitment to, and participation in, higher education reflects an awareness of science and intellectual activity promoted by the Nobel Prizes. If you look at the rest of the world, both intellectual life and democratic governance continue to be in good shape in Sweden. I wish we could say that about all the developed countries.

Winning the prize is immensely gratifying both professionally and personally, but most scientists who enjoy the extraordinary privilege of a Nobel realise it also comes with responsibilities. One is to represent their particular scientific field. Many take a much broader view and speak out regularly in defence of science, reason and justice.

I've taken advantage of several international platforms since receiving the prize to express concerns about matters like population control, global warming and environmental degradation. Like me, many Nobel laureates believe one of the major challenges we face in the twenty-first century is to deal effectively with the linked problems of environmental protection and global sustainability. Words like 'ecology', 'environmentalism' and 'sustainability' that were almost totally absent from public discourse more than forty years ago now have major political resonance. It's something Alfred Nobel could have only guessed at. Would he have instituted a Nobel Prize for Sustainability if he had lived to 1996 rather than 1896?

If we are to achieve satisfying and positive outcomes for humanity through this coming century, it is essential to bring together the different cultures that Alfred Nobel recognised by his prizes so that they interact in a mutually supportive way. Beyond that, it will be necessary to work with political and corporate power-brokers, financiers and religious leaders in the effort to build healthy and satisfying lives for all people on this small planet. How else can we ever hope to achieve harmony, peace and global prosperity?

One of the barriers to useful collaboration is that these disparate groups speak different languages. The word 'evidence', for instance, means one thing to a scientist, something else to a religious person, and something different again for a politician or a corporate advertising executive. Though everyone has some understanding of humanitarian ideals, literature, religion, politics, business and (perhaps) economics, the most obscure activity to many is what goes on in the sciences.

Many misinterpret the scope of science. The enormous successes over the past century have led to the widespread assumption that there will always be some sort of clever fix. Why shouldn't everyone drive around in gas-guzzling, polluting four-wheel drives, people ask. If the oil runs out, hydrogen will fix the problem. We can get hydrogen from hydroelectricity—otherwise, it will take either fossil fuel or nuclear power to produce hydrogen in large quantities. Strangely, although many in the broader community are suspicious of science, they are at the same time confident that science will solve every difficulty. In truth, science does some things very well, and others not so well. Look at the technologically magnificent new A380 Airbus: sophisticated electronics and control systems—and it burns oil, fossil fuel.

I strongly believe that, though modern science is highly specialised and can speak in obscure tongues, an understanding of the basic nature of scientific enquiry and what scientists bring to the communal table is both straightforward and accessible to anyone. The task of the scientist through the twenty-first century is to advance discovery, evidence-based enquiry and the technological innovation that contributes to solving problems, alleviating suffering, and generating genuine and sustainable prosperity. This can only ever be just one part of the complex, multifactorial equation that operates to improve the human condition. Though there were many disasters through the twentieth century, the advances were extraordinary. We need to continue the best part of that trajectory.

2

The Science Culture

n the movies, scientists are almost always mad, bad or quaint nerds who rattle on about controlling the world, shrinking the kids or inventing gadgets in the James Bond tradition. The caricatures may just be purely fun, or the result of discomfort or lack of understanding. Or perhaps they exist because scientific knowledge is so highly specialised, and its application has changed the world so profoundly, that people see science and scientists as hard, cold and less than human.

The basic methodology of modern science is comparatively new in terms of the way human beings have traditionally approached the world. We are used to trial and error and the common sense approach, but science works by a much more formal process that leads often to counter-intuitive conclusions. The level of insight associated with award-winning science may not be vastly different from that found in great writing, but good books are much more familiar than what we now know as science.

We all understand the social function of prize-winning literature. Stories and story telling are central to the human condition. We can look to, and learn from, the paintings and oral tradition of cultures like those of the Indigenous people who were isolated on the Australian continent for more than 40,000 years. Though the Greek historians

Thucydides and Herodotus have been dead for around 2,500 years, the translations of their works can be read with admiration, not only for the information but also for breadth of knowledge and insight they convey. The same is true for the philosophical writings of Plato, Homer's epics or the plays of Euripides. The tradition is old and strong. The mirrors our contemporary poets, playwrights and philosopher novelists hold up to us may not differ greatly in quality or clarity from previous ones, but they reflect *our* current reality and are immensely important to us.

The march of science from early times is much less well recorded. The very beginnings, from the discovery of the wheel, are lost in time. We can only speculate: were the builders of Stonehenge more than 4,000 years ago acting on carefully worked out mathematical and astronomical principles? How did they think and operate? Did those who constructed the Egyptian pyramids or the ancient cities of the Mediterranean use well-understood basic theory, or precedent and wisdom based on trial and error? The shamans, priests and witch doctors who claimed to intercede with the gods to produce rain or good crops are likely to have had a pretty random rate of success. Even if they were able to maintain a position of power through fear, ritual sacrifices and the accrual of wealth, they were also likely to be killed if chance caught up with them and they had a few bad years, and reading the entrails of dead chickens has pretty much gone out as a useful technique for weather forecasting or medical diagnosis. Contemporary scientists are neither priests nor witch doctors. The science that we live with today is, in fact, only about 500 years old.

To understand modern science, it helps to have a sense of how it developed. The earliest trace of the history of science dates from the beginning of the written word, and the available evidence tells us that science started well, but then fell on hard times. Science is neither a religion nor a substitute for religion (see chapter 7), but there was a time when the people in charge of using discovery and knowledge to promote the welfare of humankind behaved more like a priesthood than anything else.

Many who had the time and opportunity to pursue intellectual activity and enquiry were in fact priests and, as such, were constrained by religious authority. The Polish astronomer–cleric Nicolaus Copernicus (1473–1543) is said to have delayed the publication of his magnum opus on the earth orbiting the sun, *De Revolutionibus Orbium Celestium*, until just before his death from fear that he might be tried for heresy and burned at the stake. The church, back then, could be a lot more severe than modern research grant and manuscript reviewers. Still, there is other evidence that curious priests and university teachers were quietly involved in a spectrum of scientific investigations. The avoidance of boredom is, I think, an enduring characteristic of intelligent human beings. So is the search for truth.

In the ancient Greeks, we see the characteristics of what we now think of as science at its best in Archimedes of Syracuse (279–212 BC). If he were around today, he might have won several Nobel Prizes. He is credited with inventing the Archimedean screw (which is still used in parts of Egypt to raise water), the compound pulley, a variety of war machines and a spectrum of geometrical and mathematical principles. The famous story of how he jumped out of his bath and ran naked in the streets shouting '*Eureka!*' ('I have found it!') after discovering the principle that the volume of a body can be measured by the displacement of water is familiar to most. Here we see exactly how modern science often works.

Archimedes did what looked to be a simple experiment (had a bath), made an observation (the water rose), developed a hypothesis (he had displaced an equal volume of water), then reported the finding in a way that was both public (very, in his case) and intellectually accessible to anyone who wanted to repeat the study. He noticed the displacement effect because he was already thinking about a problem. The king of Syracuse, Hiero II, wanted to know if the jeweller who had made a crown for him had cheated him. Was it solid gold, or part silver? Silver is lighter than gold so it will displace a greater volume for an equivalent weight: this was clearly the way to test the purity of the crown. It is impossible to believe that nobody had ever noticed before that a full bath overflows when somebody jumps into it—Archimedes simply took something that was commonplace, thought about it, stated a basic principle of physics and presumably got paid by the king. He would also have repeated the experiment (by having another bath).

The written archives that describe the life's work of Archimedes are thought to have been lodged in the Great Library of Alexandria, which was destroyed before 700 AD, probably by fire. Though called a library, this may well have been the world's first university. Recent excavations suggest a capacity to seat as many as 5,000 students. It seems to me that the dark ages that so stultified human enquiry for a thousand years—at least in the Christian world—began symbolically with the final suppression of the Alexandria Library by the Christian emperor, Theodosius IV. The modern scientific era had to wait for the upheavals of the

Renaissance and the Reformation, and for figures like Francis Bacon (1561–1626), whose writings provided the philosophical basis of the inductive method that drives scientific enquiry today. Scientists use inductive reasoning that progresses from specific observation (data) to generalisation (theory), rather than deduction from some all-encompassing hypothetical construct. Bacon wrote that 'the only knowledge of importance to man was empirically rooted in the natural world'. We finally rediscovered what Archimedes had known more than 1,500 years previously: that nature cannot be explained simply by making statements about how we *think* it might work. Reliance on *ex cathedra* pronouncements that emphasise authority over reason, discovery and evidence is both death to the spirit of truth and enquiry and physically damaging for our species and for our world.

The long night of Galenic medicine is a good example of the danger of relying on authority rather than evidence. Galen of Pergamon lived in the second century of the Christian era. Many of his writings were based on the earlier ideas of Hippocrates and became embedded in medical practice, accepted in the same way that philosophical arguments are. Galen was the great medical practitioner of his time—among his other achievements, he was chief physician to the gladiators. Although he did some useful experiments on the kidney, by observing the effects of tying off the ureters in animals, he was completely wrong about the role of the heart. He misunderstood the circulation of the blood, believing that blood was made continuously in the heart or the liver, and then consumed in the organs after being moved around the body by the lungs. The failure of Galen and others after him to realise there is a closed circulatory system meant it was perfectly all right to bleed very sick people to remove 'bad blood'. This procedure persisted well into the nineteenth century, and is likely to have shortened many lives.

What changed in the sixteenth century is that people again started to make systematic observations and to do experiments. At the University of Padua—perhaps the first great centre for science-based medicine—the Belgian physician Andreas Vesalius (1514–1564) managed to get around the various laws that forbade dissection of the human body and wrote the first major anatomy text, *De Humani Corporis Fabrica*. While studying at that university, the English doctor William Harvey (1578–1657) began a series of experiments showing that Galen had been wrong and that the heart functioned basically as a pump, with valves that allowed unidirectional flow, enabling the blood to recirculate between the arterial and venous compartments. The discovery of the capillary circulation by Marcello Malpighi completed the circuit between artery and vein. Harvey had come to his conclusions by 1615 but, because of the power of the entrenched Galenic tradition, did not publish his seminal work, *De Motu Cordis*, until 1628. I recently saw a second edition on sale for about $US30,000.

By this time the science revolution was well established in Europe. In Paris, King Francis I established the Collège de France (1529) as a major intellectual institution that was totally independent of ecclesiastical authority. All lectures were free and open to the public. That continues today: I spoke there not long back at a scientific symposium, and any person who wished to attend was able to do so at no charge. The Collège de France remains one of

France's most powerful institutions, located close to the Institut Pasteur. The sciences section of the Académie Française was established in 1666, four years after Charles II of England granted the initial charter of what continues as the British National Academy of Science, the Royal Society of London. Every major English scientist from Newton to James Cook to Darwin and on has been a Fellow of the Royal Society.

The motto of the Royal Society is *nullius in verba*, which means 'nothing by words alone', which in turn translates to mean, 'You have to do the experiment'. The idea of experimenting is not foreign to us: children do it all the time. But experiment alone is not enough. The experimental results have to be reported publicly in such a way that anyone who has the resources to do so should be able to confirm or refute the study. It has to be repeatable. No scientific finding becomes fully legitimate until it is published and discussed. I always tell my young colleagues, 'If it isn't published, it isn't done'. The oldest scientific journal in English is *The Philosophical Transactions of the Royal Society of London*. The *Phil. Trans.* has been in continuous publication since 1665.

The seemingly simple technique of moving from hypothesis to experiment to publication and independent verification has transformed our world over the past 500 years. This change was part of a larger Western intellectual awakening which began with the European cultural Renaissance and the Protestant Reformation and culminated, in the mid-to-late eighteenth and early nineteenth centuries, in the Enlightenment. The Enlightenment in turn provided the intellectual framework for the founders of the new

American Republic, particularly for Thomas Jefferson. The questioning of received custom, tradition and authority, and the encouragement of individual reason and freedom of thought, meant that the business of explanation shifted from assertion and 'revealed truth' to evidence.

In science, knowledge can advance because each additional, small step starts from the best possible understanding of the underlying reality. Sometimes, of course, a further experiment proves to be the next *big* step that shows the earlier conclusion to have been inappropriate or incomplete. The thinking then changes, but the best scientists, like military generals, are always ready to retreat from a position that can no longer be defended, no matter how much of their personal reputation has been invested in the enterprise. I have seen a few scientists self-destruct and lose credibility in a spectacular and public way because they were unable to let go of a defunct idea or a flawed experiment. Science is not like politics: there is nothing wrong with changing one's mind when better evidence becomes available.

The big problem with building an increasingly complex body of knowledge on successive insights, experiments and discoveries is that both the technologies that are required to do the work and many of the important conclusions become less and less accessible intellectually to the non-specialist. Living in the sixteenth and seventeenth centuries, Francis Bacon and William Harvey could readily encompass the known world of science. This would also have been true for the initial group of seven who got together as the first members of what was to become the Royal Society, at Cheapside, London, in 1645 to discuss the 'new philosophy' or 'experimental philosophy'. A little later, Sir Isaac Newton (1643–1727), who famously discovered gravity from watching an apple fall, and the diarist Samuel Pepys, who was elected to the Society in 1665, would have understood everything that went on at the meetings.

Even in the nineteenth century, the British biologist T. H. Huxley was first employed in 1846 as surgeon on HMS *Rattlesnake* for its voyage to the north of Australia, then left the navy to become lecturer at the School of Mines and naturalist to the British Geological Survey. This breadth of interest is, because of the enormous increase in human knowledge, no longer characteristic of serious scientists. Incidentally, Huxley met his future wife, Henrietta Heathorn, while he was in Australia on the *Rattlesnake* expedition and, if the fledgling University of Sydney had offered him a job, the Huxley family that has been so prominent in the British scientific and literary world might have had a somewhat different history.

The difficulty now is that great scientists tend to be perceived as narrow experts—and everyone is suspicious of experts, especially if their findings often run counter to popular wishes or convenient beliefs, as can be the case with science. This tension provides fertile ground for politicians who can exploit the divergence between, say, the evidence for global warming that indicates that we must decrease carbon emissions, and the conviction of many that they can burn energy as if there's no tomorrow. Here again is the problem: the scientific community advises practical, responsible steps to ameliorate the situation, but many in the broader public believe that science will deliver miraculous answers that will absolve us from taking those responsible steps. The popular view of science and scientists is, to say the least, conflicted.

How do we foster a more sophisticated understanding of what science can and cannot do in the broader public consciousness? There are good books on science and scientists, many of which are written by dedicated science journalists. A few that I've enjoyed are listed at the end of this volume. Public radio can also do a great job; Robyn Williams of the *Science Show* and Norman Swan of the *Health Report* on the ABC are in my pantheon of Australian heroes. National Public Radio plays much the same role in the United States, and our radio dial is tuned to the local public broadcasting service station for 100 per cent of the time we spend in Memphis.

Television science programs range from the outstandingly good to the appallingly bad and, rather like the Internet medium, there may be no easy way for the non-scientist to tell the difference. The ABC-TV *Catalyst* programs that I've seen have been excellent (quite a few of them sourced from outside Australia, presumably for financial reasons). In the United States, the *Scientific American Frontiers* programs, narrated by Alan Alda, and Nova's *Elegant Universe* are consistently first-class. However, some of the worst television science efforts that I've ever seen (in various countries), dealing with important topics like childhood vaccination, mad cow disease and the origin of AIDS, were produced by organisations that one would expect to be more responsible. The problem arises, I think, when the desire to entertain and generate controversy pushes into the deep background any intent to communicate sound, verifiable information. Still, if politicians are sincere about their claims that they want to see the levels of general scientific interest and literacy improve, putting more money into public radio and television formats would be a good place to start.

Another way to bring the science culture more into the general domain is to take a leaf from the book of the Nobel Literature laureate, the Italian playwright Dario Fo, and inform via the medium of the theatre. Stage plays and intelligently made television dramas and movies (a scarce product when it comes to science) can be very powerful mechanisms for communicating. Michael Frayn's *Copenhagen* speculates about what went on at the wartime meeting between the physicists Walter Heisenberg and his former mentor and friend Niels Bohr. The central theme of this very successful play is the ethical complexities raised by Heisenberg's role in a possible Nazi nuclear weapons program, and the personal agony of the two protagonists. *Copenhagen* shows that a human dilemma created by a possible application of science can be transformed into an entertainment that is both enjoyable and intellectually challenging.

During the course of the 1996 Nobel week, Hans Wigzell, who is very personable and quick-witted, related that he was appearing regularly in a science play staged in one of Stockholm's commercial theatres. The actors had learned their lines, but didn't expect that of Hans, who simply ad-libbed whenever he was on stage. I know from my own experience that it's hard for a scientist to learn lines: we like to start from first principles and are bored by repetition. The experience was clearly great fun for Hans, and the show was evidently pretty popular. He had the sense that he was managing to get something of the essence of the scientific experience through to a broader audience, though I haven't heard that he's given up his day-time job as director of the Karolinska Institute yet.

What also needs to happen is to move science from the realm of remoteness and embed it much more in the normal human experience. Everyone can achieve at least a measure of scientific literacy and, what is more, people will gain both personal satisfaction and even a sense of wonder and delight from having a better understanding of the natural world around them. Science teachers are enormously important people and we should value them, and pay them, highly. Field trips, lab experiments and so forth can give children a sense of what science is like, but the tendency, naturally, is to set up demonstrations that work. Perhaps it would also be useful to develop ways of conveying the sense that a measure of frustration and uncertainty is also a normal part of the reality with innovative science. There can be a worm in the shiniest apple and sometimes, of course, the worm is more interesting than the apple. Do kids get bored with school science because they don't see that intriguing worm and, instead, come away with the impression that the whole game is predictable and mechanistic? Nothing, of course, could be further from the truth.

Science museums with creatively designed interactive exhibits are found in most major cities. Visiting them is an experience that both children and adults can enjoy greatly, as we did when we lived in Philadelphia—our kids had a great time in the Saturday morning science program at the venerable Franklin Institute. Increasingly, as is already the case with the Nobel e-museum, it will be possible to visit on the Web reconstructions of some of the experiments that led to major discoveries.

In general, scientists need to develop a better understanding of how to communicate with both the general public and with political leaders. Many have the naïve impression that the world outside their laboratory is poised on tenterhooks to hear their latest, authoritative pronouncement.

Television and the print media provide obvious avenues of contact, but any scientist who has tried to use these means soon realises that journalists are professionals too and, though they will react positively to a good story, are not there just to be used. The economic realities are also such that far fewer newspapers have professional science journalists on their staff than they did ten years ago. We were particularly spoiled at St Jude because the *Memphis Commercial Appeal*, a Scripps/Howard paper, employed Mary Powers, an outstanding science reporter. Being a kids' cancer research hospital with high quality basic science research, St Jude provides regular, good science stories that have an additional, human-interest component. In general, scientific institutions should do everything possible to develop open, amiable and honest lines of communication with local media outlets. In addition, course work and practical experience in science communication should, I believe, be included in every PhD program. Young scientists enjoy talking with primary and high school students, both about what they are doing in the laboratory and their experience of the scientific life in general. It shouldn't hurt to tell senior high school kids that serious young scientists go to meetings at ski resorts and are also party animals!

What about focusing more on convincing the key decision-makers? Politicians respond to popular pressure, so evidence of broad concern can help to raise the profile of a particular issue. Lobby groups can work well in the United States where every member of the House and Senate acts as an essentially free agent. This approach is not as effective in the more disciplined structure of a Westminster-style parliamentary democracy in which back-benchers have relatively little autonomy. Some political systems, such as Canada's with its recent Health Research Institutes initiative, take science very seriously and establish strong bureaucratic structures and lines of communication with the research community to provide specialist advice.

Both the British and the French governments make heavy use of their national science academies, the Royal Society and the Académie Française respectively. The US National Academy of Sciences has traditionally played a major role in providing well-researched, dispassionate and detailed advice to both the Congress and the President, the job that it was set up to do at the time of its foundation by Abraham Lincoln. There is nothing to stop a US president drafting a top scientist for a Cabinet post. This happens also in Taiwan, which has copied the US model of an appointed rather than an elected Cabinet. Each major British government department has a chief scientist and associated office. In any given parliament a few members of the House of Commons will have some scientific background, and eminent scientists and medical professionals are quite often appointed to the House of Lords.

One problem in the scientist–politician relation is that the two cultures are so different. Most politicians in any system trained first as lawyers. Good lawyers are great debaters, and have a tremendous track record of winning arguments. Being a great debater doesn't help much, though, if someone who has been duly counselled finds himself alone and unarmed on a narrow path facing a huge, hungry, brown bear. Scientists can issue warnings about the particular big brown bear they see ahead for society, but politicians may either take no notice or choose to hear the guys who argue that this bear isn't really dangerous and, in any case, we need the bears for tourism. Though scientists cannot and should not dictate policy, it doesn't take much insight to realise that every major political decision should, where applicable, be informed by the best possible scientific insights.

Some groups in society are simply unlikely to ever be sympathetic to science, even though they may have no qualms about taking advantage of what science and technology have to offer. Perhaps the biggest disconnect is in the minds of those at the extremes of religious fundamentalism who, if their wishes were implemented, would return to social models that depend on 'revealed' truth and authority. Science and discovery, by comparison, depend on openness to new ideas, the willingness to follow processes to their logical conclusions and flexibility of action and movement, all of which are antithetical to most fundamentalists. Many first-class scientists are women with children who would, of course, either be denied the opportunity to work or have their careers compromised in some fundamentalist scenarios.

The contradiction is that people in such groups are often sophisticated in other ways, and, for instance, use effectively tools provided by modern science and technology. Techniques of mass persuasion operating via television, radio and the Internet are exploited to spread their beliefs and to raise funds. The voice of the preacher at an evangelical meeting may be manipulated to a God-like resonance by electronic distortion and powerful amplification, while his enormously enlarged and enhanced visage dominates from one of those massive display screens otherwise encountered only in sports stadiums and airports.

Though many religious fundamentalists are decent people, motivated by a sense of both cultural disintegration and the desire for a moral and safe upbringing for children, it may be difficult to establish a meaningful dialogue between their world view, based on authority and doctrine, with the process of continuing revolution that is characteristic of contemporary science. The so-called conservative political parties play a dangerous game when they foster fundamentalist agendas in order to capture a solid voting bloc. The growth of knowledge-based economies depends on flexibility, creativity and the capacity to retain those with the greatest abilities and the best minds, no matter what their belief system, sex or sexual orientation may be. Regressive social models quickly drive those people away, or even destroy them.

Fundamentalism is not, of course, confined to religious groups. The extremes of the environmental movement are absolutely opposed to the application of genetic engineering approaches to improve plant varieties. Nothing in the human experience provides more fertile ground for faddism than the food we eat. Though traditional plant breeding has changed most of the items in our food basket to the point that they would be unrecognisable by those living in the fourteenth century, the thought that techniques based in contemporary gene technology should be used for further, targeted improvement is apparently unacceptable to many environmental activists.

Part of the problem is that the first applications of plant gene technology were fostered by a large chemical company, Monsanto, in a way that horrified those who believed that plant varieties, seed supplies and so forth should not be controlled by international conglomerates. Many entirely reasonable people share this perception, including most of the Aid agencies operating in the developing world. There were few, if any, regulatory controls, and the scientists didn't help by simply telling people that there was no danger, and that they should just be allowed to get on with it. The trust-us-and-accept-what-you're-told approach doesn't work too well anywhere any more. A more judicious approach that invites openness and accountability would no doubt serve both the public and the scientific interests much better.

The reaction against genetically modified organisms (GMOs) has generally been greater in Europe than in the United States. Part of the reason is the relative strength of the environmental movements. A more cynical view is that Europe sits on a mountain of food as a consequence of the EU's common agricultural policy. One way to prevent competition from the highly efficient agriculture sector in the United States, or from low-cost producers in Africa, is to ensure that public opinion is resolutely opposed to GM foods. Any African country that follows the US example of adopting GM approaches to help feed its own livestock and people would, like the United States, be blocked from exporting plant products to Europe. A further component that feeds into the European discomfort is that there is a general and justifiable distrust of regulatory authorities. In part, this may stem from, for example, the way that the then British government and bureaucracy handled the BSE (mad cow disease) outbreak. Americans have much more confidence in agencies like their FDA, which controls the approval of drugs and vaccines for human use.

When I was visiting England and staying in hotels in the mid-1990s, I often poured soy milk from the breakfast buffet on my corn flakes. Soy milk, an excellent product that is generally made from GM soy, is good for someone like me with cholesterol problems, but it disappeared from English hotels and supermarkets when the furore about GMOs broke in the media. The negative reaction has also compromised the distribution of products like saffron rice, which is freely available and was engineered in Swiss public sector science laboratories to correct the blindness caused by vitamin A deficiency and the iron-deficiency anaemia that afflicts people, particularly women, in some parts of the developing world. We recently saw the tragedy of Zambia refusing GM corn, which has been eaten by millions of Americans without any untoward consequences, in the face of a disastrous situation where many were starving. Not only is there a complete absence of any evidence that GM corn or saffron rice has any deleterious effect, there is no good scientific reason either for thinking that there could be such effects.

Every time I talk to plant scientists, whether they work in the advanced nations or in developing countries, I get the same message: they want to use GM approaches for applications like the production of higher yield crops, disease resistant varieties and plants that grow on poor soils, especially those with a high salt content. The same is true for the Consultative Group for Agriculture Research (CGIAR), the organisation started in 1971 by the former US Secretary of Defence and president of the World Bank, Robert McNamara, and the Australian economist Sir John Crawford, for channelling international resources to science directed at feeding the poor of the world. The CGIAR currently operates some twelve agricultural research institutes in developing countries. In the past, they funded the research done by the maize breeder Norman Borlaug that led to his Nobel Peace Prize for contributions to the 'Green revolution' of the 1960s. Borlaug is an energetic, enthusiastic and infinitely decent human being, now in his eighties, who is still doing everything in his power to promote science-based agriculture as a partial solution to the joint problems of starvation and social degradation. He was a conventional plant breeder and is a passionate advocate for the use of GM approaches to speed up what needs to be done.

An easy way to access an indigenous African plant scientist's perspective on the GMO debate is to go to Google and type in the name Florence Wambugu. In her address to the US Congress on the use of GM approaches to enhance food production in developing countries, Florence argued that the primary achievement of the European anti-GM lobby has been to 'keep safe and nutritious food out of the hands of starving people'. She doesn't pull any punches and attracts equally virulent responses. A Google search will also provide a spectrum of positive and negative assessments of Florence's views. Reading some of those comments gives a pretty good idea of just how polarised this issue has become. You will see attacks on her ethical values, her character and her science, most of which are, I assume, written by well-fed people sitting comfortably in the northern hemisphere.

The fact of the matter is that at least 800 million people of the 6.4 billion or so that inhabit this planet do not get sufficient to eat each day. Aid certainly helps, but moving food internationally requires ships, trucks and the associated consumption of energy. Care has to be taken to ensure that what does cross the seas isn't stolen to end up being sold to starving people. If that food transfer strategy worked, however, there is no reason—given the global surpluses—why the problem should not already have been solved. Nobody and no country wants to live as a charity case. The ultimate aim must be to ensure local sustainability in food supply, whether at village or national level. Pride and psychological well-being are at least as important as efficiency when it comes to the food/agriculture equation. The GMO approach is certainly not the whole solution, but it does have the potential to play an important part.

What is to be done? Open, rational public dialogue would help, but that's difficult at the moment. Some organisations have made opposition to GMOs an article of faith, and so many individuals assess the issue in a very emotional way. There are, however, a number of obvious steps to take. The first necessity is to ensure that the types of regulatory controls and monitoring procedures that elicit public confidence are not only in place but can be seen to be working. The second is to strengthen public sector biotechnology research and development, in Africa particularly, so that the countries themselves have 'ownership' of both the science and the applications of the science. The third is to build economic structures that make GM seeds available free, or at cost, to farmers at the village level. There is no reason why large-scale agriculture operations in Africa or anywhere else should not pay commercial rates. As Florence Wambugu points out, there is nothing to stop farmers who plant GMOs from also maintaining their own, conventional seed stocks produced in traditional ways. It must be obvious that GM plant varieties will be used in the first place only if they are associated with reduced fertiliser and insecticide costs, or give substantially higher yields.

The irony is that many scientists who support GMO approaches are also passionate advocates of environmental conservation. Why should strategies that have the potential to bind marginal soils together, or limit the use of insecticides and nitrate fertilisers, be considered 'anti-environment'? There are enough really bad guys out there in positions of power without having those who are concerned with promoting human well-being and the health of the planet taking opposite sides on this important issue. It is surely time to initiate a more sensible and balanced discussion.

Unlike the situation with plant GMOs, people in general have relatively few problems with modern medical science. Everyone is happy to benefit from a new cancer cure and there is no doubt in anyone's mind that a more effective treatment is a real plus. The next generation influenza vaccines are likely to be GMOs engineered by a process called reverse genetics. I doubt there will be any problem with their acceptance in the face of an influenza pandemic, especially if, like the H5N1 bird virus that is a looming threat (see chapter 4), there are indications that it may cause severe infections and human deaths on a massive scale.

Perhaps the main area of debate relates to the use of embryonic stem cells, where the lobby groups termed 'pro life' (the life of the foetus, not the woman) condemn elective or medically advised abortion in the belief that success with stem cell research will in some way legitimise the 'pro choice' (by the mother and doctor) position. This debate is a social and moral one that will not be swayed by scientific argument. Apart from deeply held religious convictions, there is also a broader cause for concern that any widespread use of early human embryos from terminated pregnancies could become a means of providing 'spare parts'. Ask yourself whether some of the totalitarian monsters that so blighted the twentieth century would have had the slightest qualm about paying (or forcing) a young woman to carry a foetus to three or four months if the tissue from that embryo would have ensured a further year of life for them. Might there just possibly be people like that around today? This has to be a broad debate that is informed by both science and ethics. Fortunately, the great majority of medical science research has nothing to do with embryonic stem cells.

Few would object if, for example, we could take 'self' stem cells from an individual's own blood or bone marrow, then manipulate them in some way so that they return to the embryonic state. Such cloned, pluripotent cells would be of immense medical benefit to the individual providing them. In addition, this strategy would circumvent the difficulty with tissues derived from aborted foetuses, which bear foreign transplantation molecules that can promote rejection (see chapter 4). As with a transplanted kidney, such host-versus-graft responses can be controlled with suitable immunosuppressive drugs, but the situation is not ideal. An alternative, and utterly abhorrent, possibility would be to use 'identical twin' cells from an embryonic 'clone' of the person to be treated. This is one of many reasons why the cloning of whole human beings should be banned permanently. At the same time, great care has to be taken that any anti-cloning legislation is not drafted so broadly that it would inhibit the useful application of cloned cell lines and so forth that I've mentioned above.

The interaction between science and government is both complex and long-term. Early examples relate to weapons technology and fortifications. Archimedes designed effective war machines. René Descartes (1596–1650), who is regarded by many as the first modern mathematician, earned his living as a military engineer. The publicly funded voyages of exploration from Europe from the sixteenth century to the eighteenth were concerned both with acquiring territory and with discovering precious metals, plant varieties and so forth that could be exploited for economic advantage. Early successes fostered by this approach included the importation of tobacco, the tomato and the potato, which comes from the South American Andes.

The 1769 voyage of the His Majesty's Bark *Endeavour* under Royal Navy Lieutenant James Cook was a Royal Society scientific expedition that sailed to the Pacific to map the transit of Venus. On the return voyage, Cook was given the further task of investigating the east coast of the partially mapped land known as *Terra Australis*. The *Endeavour* also carried the naturalists Joseph Banks and Daniel Solander, who collected and described the flora and fauna of the new territories. On their return to England, the novel plants were cultivated at the Royal Botanic Gardens at Kew. Sir Joseph Banks went on to become the longest serving president of the Royal Society and it was he who suggested the 1788 establishment of a penal colony in New South Wales, which marks the beginning of European Australia. Modern Australia may be the only country on earth that was established on the suggestion of a scientist as a result of a scientific expedition.

There are many stories of how government-funded science has served the public interest, a well-known example being the development of radar which was so instrumental in the defeat in World War II of the short-lived Nazi empire. More recently, when the New York philanthropist Mary Lasker started to raise funds to combat cancer, she quickly realised that the amounts required could never be found in the charity sector. Following the advice of Senator Jacob Javitz, she persuaded President Nixon to initiate the so-called war on cancer that led to the establishment of the US National Cancer Institute (NCI). The annual budget of the NCI in 1971 was about $US180 million. It is now in excess of $4.5 billion. Apart from the advances in the understanding and treatment of cancer (particularly child-hood cancer) that have been achieved, the NCI Special Virus Cancer Program of the 1970s also helped put in place the technology that allowed the relatively rapid iso-lation of the human immunodeficiency virus (HIV) that causes AIDS.

If AIDS had hit the Western world a hundred, or even fifty, years earlier, the confusion and fear about what was happening would have been infinitely greater than it was, as there is no way that we could have isolated the causative agent using the available methods. Even after the virus was identified, some still argued for a time that HIV did not cause AIDS, a silly and irresponsible debate that cost many lives in the poorer countries. Without being able to test for the footprints of the virus, the blood supply would have remained unsafe and the social backlash against those suffering from the disease could have been much worse. The history of witch-burning, killing minorities and so forth that accompanied the disastrous plagues of the Middle Ages in Europe might well have been repeated in modern guise if AIDS had continued to spread unchecked in the democracies.

In general, democratic governments from both the liberal and conservative ends of the political spectrum have no philosophical problem when it comes to supporting medical research. Like everyone else, politicians grow older and become increasingly aware of their own health and ultimate mortality. The expense of medical research is infinitely less than the cost of health care delivery, which is now consuming such a high proportion of national budgets in every advanced country. Research scientists also plough any grant money back into the economy by buying expensive chemicals, isotopes, plastics and sophisticated equipment from private industry sources.

The discoveries made by the publicly funded biomedical research community flow on to the private sector in the development of new biotechnology start-ups and larger innovative companies. The drug industry, in particular, depends heavily on this transfer of novel findings from the public to the private world as a source of new product development. Major pharmaceutical companies set up their own laboratory complexes close to, or even inside, the top research institutes and universities. Such partnerships make good economic sense, as it is only the private sector that can ultimately assemble the resources needed to bring a new drug or vaccine to the market.

The tension between science and government comes to the fore when the best advice that the scientists can provide is seen by politicians as having acute, negative consequences on the economic (and thus the political) front. The classic case at the moment concerns the predictions of progressively escalating temperatures, melting ice caps, raised sea levels and disrupted weather patterns as a consequence of increased carbon dioxide levels. At least in the United States and Australia, the oil companies, the mining companies, the timber-cutters, the automobile industry and others do not want even to hear about global warming. On the other hand, much the same sensitivities that influenced the GMO debate have led to an acute awareness of the issue in European countries. Britain's Prime Minister Tony Blair is clearly taking global warming very seriously. Some suggest that even as early as 2030 the massive flood barrier that protects London could be rendered essentially ineffective. Everyone knows about the little Dutch boy and the dyke: it could be that his job will become much harder.

The global warming equation is, of course, extremely complex. Though it isn't my field, my impression is that the data being obtained from the ice cores, the conditions in the deep oceans and the changes in areas like the Canadian tundras provide pretty compelling evidence that we are going through a period of consistent warming following a rapid rise in carbon dioxide levels that began with the industrial revolution. According to the Chemistry Nobel laureate Sherwood Rowland, the sequential ice-core sampling that has been done from some of the glaciers has had to be discontinued because the glaciers themselves have melted. The effects may, at least initially, seem paradoxical. Melting of the ice caps is likely to result in the loss of the warm Gulf Stream that gives Britain and northwestern Europe a much milder climate than would otherwise be the case. Some argue that these areas would, for a time, enter another ice age. The southwestern United States, on the other hand, would rapidly become hotter.

None of this is set in stone and even those who believe strongly in global warming suggest varied scenarios. Given the widely expressed doubts that come, in the main, directly or indirectly from industry sources, it might be thought that governments would want to increase the amount and depth of the research being undertaken so that better predictions can be made. The telling fact is that the administrations most opposed to the global warming idea are actually cutting funds for both environmental research and the enforcement of regulatory frameworks that mandate clean air and so forth. This is certainly not the way that any scientist would approach the problem. Where there is doubt, more experiments need to be done, more observations must be made. Whether or not that is the intent, any moves to inhibit such enquiry suggest a deliberate strategy to suppress the truth: as Bill Clinton might have said, 'This dog just won't hunt'. There is too much concern, and too many players involved in too many countries.

If there were even the hint of a military threat as serious as that being put forward by the informed science community about the consequences of rapid climate change, the response by governments would be both immediate and dramatic. Why is there such a difference? A facile argument is that military spending is a traditional way of transferring public money to powerful corporations, while at the same time facilitating the type of pork barrel politics that allows local representatives to bring home the bacon. This dynamic is yet to emerge in the environmentally friendly industry sector, but there is no rational reason why moving towards the development of sustainable technologies should not be just as good a mechanism for spending tax dollars to support private sector job creation.

There are also, I think, deeper reasons that rest in the biology of what we are. Human beings are programmed by their evolutionary history to react very rapidly to the possibility of imminent attack. There was no problem, for example, in getting the global community to work together to limit the SARS epidemic. The response was universal, rapid and efficient, partly because it built on the global influenza surveillance network that operates out of the World Health Organization (WHO) in Geneva. The existence of this WHO program reflects that influenza is a continuing, immediate threat, with new strains emerging constantly. When it comes to insidious, long-term environmental damage, though, we are much less concerned about what is happening. The worm has turned now, but many agricultural communities in the developed world came very late to measures that would minimise erosion, the saltation of soils and so forth. Deforestation continues globally at alarming rates.

Protecting established industries may make economic sense for governments that are concerned only with the present, but what if this is at the expense of long-term economic vitality? While the Detroit auto-makers have been persuading normal American suburbanites to drive around in modified light trucks and military vehicles, Toyota and Honda have been marketing gas/electric hybrids that are both fuel-efficient and minimally polluting. Europe has long focused on producing more economical cars. Even without global warming, which strategy is likely to be more productive as oil supplies dwindle and/or become more expensive? Both the current political situation in the Middle East and the consequent massive cost to the US taxpayer may be considered clear consequences of an over-dependence on oil. It makes no sense to be locked in to old, dumb technology.

No matter what the future, energy independence and technologies that are clean, green and minimise waste are clearly the way to go. Why isn't this seen as a major priority? Is it because entrepreneurs operating on a small scale in what will initially be high-risk ventures do not constitute a significant source of political contributions? My bet is that those nations that focus on environmentally friendly, innovative industrial development will be the economic powerhouses of the future. The scientists, the economists, the industrialists and the politicians need to be in a process of continuing dialogue. Both the carrot of R&D funding and the stick of regulatory requirements can help to lead the innovation donkey to water!

At some stage there inevitably comes a tipping point where the basic realities change for the worse and political leaders who have allowed a disastrous situation to develop pay a severe penalty. This tends to happen very suddenly. Living in Philadelphia through the transient oil crisis that occurred in the latter part of the Carter administration, we tried to buy a Honda Civic, the smallest and most fuel-efficient car available. In the words of Lane the Butler in *The Importance of Being Earnest*, there were none to be had, 'not even for ready money'. The first sign of a tipping point with global warming may be starting to emerge in what look to be extremely atypical weather patterns. Florida experienced four substantial hurricanes last year. The minimum dollar cost is estimated (in early 2005) at 42 billion, with at least 20.5 billion being paid out as insurance compensation. As a consequence, household insurance premiums have increased dramatically and some are unable to secure appropriate coverage. This type of financial dynamic has a way of convincing people in the street that something very substantial is happening.

Education is clearly the key if we are to build a greater appreciation of science into both the political and the public consciousness. People don't have to know the details of this or that scientific discipline, but it is important to develop a reasonable understanding of how science works and where the strength and limitations of the scientific approach lie. The scientists themselves need to be better versed in the business of science communication, and there is a requirement for more, full-time professionals trained as communicators. I have already referred to the role of serious science journalism and the media, which of course includes the Internet; the difference is that we use the Net to search and discover, rather than for taking in information passively. I would, however, reiterate that innovative approaches like science theatre and the creative use of the Internet merit further development and thought.

As the Nobel physicist Richard Feynman put it: 'For a successful technology, reality must take precedence over public relations, for Nature cannot be fooled'. Science is the best tool we have for not fooling ourselves and for probing the reality of what faces us. Humanity will ultimately pay a heavy price if those of us alive today choose to inhabit a self-serving fantasy world that denies reality.

3

This Scientific Life

The day a young person starts work in a laboratory as a graduate (postgraduate) student is the day he or she joins their particular international research community. Although they may work in one building on a particular campus in a particular country, scientists operate across boundaries and in an international culture. They travel between states and continents for employment and as participating members of global communities. International journals publish research results and scientists at all levels contribute to publication, as well as travelling to key international meetings and symposia. Young scientists quickly learn that the values of science are universal. Still, most graduate students will spend some years tied to one place as they progress through further study and research, doing their 'apprenticeship' on someone else's team. A typical stay lasts three to five years. At the earliest stage, a research career rarely looks like the obvious road to riches but, because a vibrant science culture is increasingly seen to be in the national interest, there are well-established support systems, studentships, scholarships and so forth for beginning investigators. Their path is generally a lot easier than that for young academics in, say, the humanities.

Despite our fantasies of being able to work in a cabin overlooking the beach, the experimental scientist, no matter how successful, is never completely liberated from the regulated, institutional setting and the busy laboratory. The nature of the work means experimentalists are generally tied to machines and institutions. Even theoretical physicists, who may once have lived happily in the equivalent of a garden shed with a slide rule, pencil and paper, now depend mostly on the type of high-powered computing that can be accessed only from a well-resourced research institute, university or a shared national resource. The 2004 Physics Nobel, for instance, for 'a colourful discovery in the world of quarks' went to the theoretical physicists David Gross, David Politzer and Frank Wilczek from the University of California at Santa Barbara, the California Institute of Technology and MIT respectively, institutions unlikely to put their top scientists in garden sheds. Of course some top medical research places, like the Salk Institute and the Scripps Research Institute, both in La Jolla, California, do have offices that (at least for the senior people) look out over the sea, but many are located in the inner cities, close to the older medical schools.

The garden shed model is also out because scientists of all backgrounds are gregarious creatures, and use each other as sounding boards. The theoreticians I know need to talk a lot with at least one close colleague who understands intimately what they are saying, and even the purest of pure mathematicians, who may be reluctant to reveal what they're thinking to the person in the next office, tend to gather at places like the Princeton Institute of Advanced Studies. The life of John Nash, an applied mathematician from Princeton who won the 1994 Economics Nobel— and gave his name to the 'Nash equilibrium' that is central to game theory—was portrayed recently in a Hollywood film, which tells the story of the support he received during the course of his long struggle with schizophrenia, which has now achieved a great measure of remission.

The impact of mathematically based theory on the biomedical research community is a relatively recent phenomenon. The theoreticians who develop mathematical models of biological systems are generally keen to get together with experimentalists like me so they can access real data—but I fear we often disappoint them. The process of generating the numbers they would like to have could consume our whole lives and millions of dollars. Even so, theoreticians are an amiable bunch: they tend to drink a lot of coffee and beer, wave their hands around and write equations on white boards. Sometimes it is even possible to grasp what they are on about. Among others, we've worked and published with Martin Nowak, the Vienna, Oxford, Princeton and now Harvard theoretician who, incidentally, tells lots of good jokes.

Martin started as a mathematician in Vienna, then began the biology-related part of his career with Bob May, a Sydney University-trained theoretical physicist who progressed via Princeton, then Oxford, to become the Chief Scientific Advisor to the United Kingdom government. Martin applied the predator/prey relationship models that Bob had developed during his time as professor of zoology at Oxford to analyse the dynamics of the virus/host interaction in AIDS, a study that many of us found to be extremely informative. Anyone who is deeply interested can read their book, *Virus Dynamics* (published, of course, by Oxford University Press). Now Lord May, Bob is the 58th president of the Royal Society, the second Australian to serve in that role (the other was Howard Florey), and

Martin heads a new program in evolutionary dynamics at Harvard. Scientific careers don't run on rails and outcomes can be both unpredictable and extraordinary.

The cultures in the various areas of experimental science differ greatly, though we all talk a language based on hypothesis, experiment, observation, open reporting and independent verification. The scale of the 'small science' done by most biologists can, however, seem a long way from the world of the experimental physicists, what many scientists think of as 'the big end of town'. The new CERN particle accelerator located on the border of France and Switzerland cost $US2.5 billion, but will provide a unique international facility to address the most fundamental questions in physics. When the CERN people talk about an experiment, they may be describing a commitment that involves years of construction, preliminary testing and preparation. A single 'big science' experiment can cost millions—especially if it requires shooting things into space—and involve hundreds of people. Many university astronomers and physicists, though, work on the smaller scale that is familiar to a biomedical scientist like me. However, even if the world of cosmic scientists probing the beginning of the universe seems very different from that of the research biologist, the common link is that we all come down at one point or another to dealing with new data.

Laboratory scientists work in groups. A typical group of ten or twelve will be led by the 'principal investigator', or PI, the senior scientist responsible for raising the funds, setting the overall direction of the research program, seeing that the primary papers and review articles get written, and travelling to talk about the work at scientific meetings and research seminars. Every major university and research institute has a regular seminar series and, in a big scientific environment like the United States, any PI who is doing substantial and interesting research will be on one circuit or another. The cachet of a Nobel Prize, of course, only adds to this load. The more successful people are often booked for a year or more ahead, and are also solicited regularly to write review articles or chapters for books of collected papers.

The key resource person in any sizeable research program is likely to be the head technologist. Typically, this is a very experienced and competent individual with an undergraduate or master's degree who has always worked under supervision at the hands-on end of research. My Memphis laboratory is managed by Elvia Olivas, who grew up in Mexico, trained in science, worked as a technician and then took a business degree. Sometimes a PhD scientist who has decided not to pursue an independent career fills the head technologist role. Their job is to make sure the laboratory runs properly by ordering equipment and research supplies, supervising junior technologists and developing new technical approaches. The head technologist will often look after the budget, no small responsibility in any lab. Experimental biology, for instance, is no longer done solely with small items like test tubes and petri dishes, though buying plastic plates and flasks does consume a lot of the budget in many laboratories.

Even the 'small science' pursued by medical research workers commonly costs between $US600,000 and $US1.2 million a year per laboratory, including salaries, if it is to run at a competitive level. A further major expense is the price of various types of counters and analytical instruments; such resources will generally be shared between groups, but they come in the range of hundreds of thousands of dollars and have to be regularly upgraded or replaced. If a competitor can measure something new and different, or get the same result more quickly with a novel piece of equipment, a program will rapidly fall behind. Equipment generally comes under the heading of infrastructure costs and, as machines aren't particularly glamorous—unless they're so impressive physically (like a Jupiter rocket) that politicians can put them on display—they are often a good target for philanthropists who want to help the medical research enterprise.

Some institutions, like those heavily into electrical neuroscience, have traditionally maintained their own workshops with highly skilled engineers and craftsmen, glass-blowers, electronics technicians and fitters who build one-off items of equipment. In the days when ships and sea travel were still the norm, constructing equipment was an important activity for competitive scientists in isolated countries like Australia. Some of these ancient machines are still around.

A young neuroscientist, Michael Pender, who worked in the department I ran during the 1980s at the John Curtin School of Medical Research (JCSMR) in Canberra, was given some old recording devices that had been in storage for years. They functioned for a time, then went up in a cloud of vile, choking black smoke. This was tough on Mike because, as he reminded me years later, the storeroom that had been converted to provide him with laboratory space was very poorly ventilated. In fact, when the door was shut, the incoming air supply was via an exhaust fan from the men's bathroom. I doubt that Occupational Health and Safety people would allow that today.

At least in laboratories in the United States, junior technologists are likely to be learning the science trade while earning some money as they apply for entry to graduate school or medical school. It's a good training for someone who eventually ends up as a medical doctor. The small immunology department I headed for fourteen years at St Jude sent at least four such young people off to medical school, mostly to the University of Tennessee but one also to Puerto Rico. A few who ultimately qualify as MDs may come back to research. The other members of any given research group will consist of PhD students and post-doctoral fellows (postdocs), with greater numbers of post-docs in the high-profile research universities and institutes. Individuals in both groups will have in mind the goal of becoming independent scientists in academia, though a few may already be aiming for a job in a science-based industry setting. Most, if not all, of the postdocs will have come to the laboratory after completing their PhD work elsewhere.

Over the years, I have had young colleagues in my various laboratories from Australia, the United States, Scotland, Hungary, Switzerland, Iran, England, Mexico, South Africa, India, Italy, New Zealand, Poland, Korea, Denmark, China, France and Nicaragua. This is a fairly modest listing, as I don't run a very big program. I can count four MDs, two veterinary scientists and the rest were technicians, graduate students and PhD scientists. Two of the MDs (Neil Greenspan and David Schwartz, both Harvard undergraduates) were in the process of completing the joint MD/PhD in the medical scientist's training (MSTP) program at the University of Pennsylvania. The MSTP is a US scheme to attract young medical doctors into science by offering the incentive of a reasonable living allowance and the payment of all fees. Otherwise, a big problem for MDs who want to pursue a less well-paid research career can be the massive financial debts that they incur while training. Two Nobel medicine laureates, Al Gilman (1994) and Ferid Murad (1998), graduated from the early MSTP established by Earl Sutherland (Medicine, 1971) at Case Western Reserve University in Cleveland, Ohio.

It's at the postdoctoral stage that many young people make their first international move. Depending on the society they come from, they may or may not be aiming to go back. For those from less well-resourced countries, the postdoc transition is often 'the great escape'. It is an extraordinary but true fact that many societies, even some that are quite wealthy, are led by politicians who provide such limited support for science that they effectively export their nation's best and brightest. The universities and science-based industries of Western Europe and the United States have been the great beneficiaries.

Australia has also benefited from an influx of young Asian students, who come from the countries immediately to the north to train in the leading universities and research institutes. At the same time, many of Australia's own young stars have ended up living permanently in the northern hemisphere. There is a measure of predictability about this for a scientific culture in a country with a relatively small population. The brightest undergraduates will want to do their PhD work with the top scientists who, inevitably, bring them up to be quite a bit like themselves. Beyond a certain number, such people can be absorbed only internationally, not locally. The best advice for those who go overseas for postdoctoral training and want to come back to their home turf is to learn different skills and approaches that will allow them to establish a new, independent area of activity on return. Still, though no nation should be complacent about losing top talent, it is important to recognise that there will always be a 'global churn' when it comes to the movement profiles of innovative scientists. Controlling the migrations of such people is a bit like herding cats, a rather useless exercise. It does nonetheless help to offer quality cat food (funding) and plenty of it if you want to attract the attention of a top cat.

The decision about where to continue the scientific apprenticeship begun as a graduate student can be the most important in any scientific career. Though it may look overwhelmingly attractive to go as a postdoc to the laboratory of the person who seems to be at the cutting edge of the subject, a lot of other, bright young people will have the same idea. Sometimes it can be hard to get your work noticed: one colleague I like and respect greatly for his scientific achievements told me he doesn't even bother to talk with postdocs who aren't generating interesting data. This is fine for highly competitive and aggressive young people, but it doesn't suit others who may ultimately turn out to be very good scientists. The best way for an intending postdoc to get the feel of a program is to visit and talk with the junior members of the laboratory, preferably over a coffee or a beer. It can often be a good experience to work with a younger PI who is still on the way up, runs a smaller operation, travels less and is able to give more personal attention.

All leading PIs know that they will be only as effective as the young people who work with them. They are the ones who do the hands-on experiments, look first at the results and process the data sets that are the life-blood of experimental science. What comes out in the end analysis is the product of an intense and continuing dialogue between the members of the group. Different PIs operate in different ways, but most encourage open discussion and debate. Classically, the others in the group may first see a new set of results and hear the possible interpretation at a weekly laboratory meeting that all are required to attend. They can then add their own ideas about what a particular finding may mean in the broader sense, or what could happen next. In a dynamic program all ideas are up for discussion, whether they come from the PI or the newest graduate student. Sometimes fresh, young minds that are not loaded down with what went before will see a particular set of data, or a possible opportunity, from a new and interesting perspective.

Misty Jenkins, a Koori scholar who traces part of her genetic heritage back to Australia's Indigenous people, is a graduate student in my Melbourne program. She is working on influenza virus-specific 'killer' T cells, the assassins of the immune system which I discuss at greater length in chapter 4. She was recently at the Australian Society of Immunology meeting in Adelaide when she chanced to talk with another young scientist who is using digital imaging approaches to analyse visually how immune cells kill. Now she will get together with him to take a look at some of the processes that we have been studying, using an entirely different approach, one that nobody else in the laboratory, including me, had been thinking about. Will it tell us anything new? We don't know yet, but it sounds intriguing and that's often the way that something different and exciting gets off the ground.

The postdocs and graduate students in any good laboratory are generally driven by an intense sense of excitement and intellectual curiosity. Science at its best is, in the end, like a good detective story. It offers the chance to ask the questions, search for clues, uncover what is hidden and come to a conclusion or, even better, a solution. The main word for the scientist at every level is 'interrogate'—and most Nobel Prizes go to the 'hard' scientists who interrogate nature itself. The classical interrogation technique of the scientist is the experiment and, as with any detection task, the first step is to ask the right question.

The key insight about the right question might come initially from reading or talking with theoreticians who think full-time but don't do experiments—rather like the paralysed detective, Lincoln Ryme, in Jeffrey Deaver's crime novels, who is all intellect and no praxis. As the investigation progresses, more questions will be provoked by what is discovered, and they will drive successive experiments. Like any human activity that seeks to intrigue and inform, the whole task is concerned with putting together a good story. If the team is both lucky and skilled, the ultimate account will be novel and interesting. Often, though, despite a promising start, the conclusions can be mundane and boring, and many of the resulting research papers are at the level of 'not proven' (a verdict that is possible in the Scottish legal system, an alternative to the 'guilty or not guilty' that is more familiar to most of us).

If the experiment does lead to a good result, it has to be confirmed before we can accept that it can indeed be believed. The experiment may be repeated exactly as before, or changes could be made that might fill in a few gaps in the story while still providing support for the initial analysis. Sometimes an experimental result can look just too good to be true, and a repeat study will bring us back to Earth. What has generally made me very comfortable about the results I see from my research group is that the data sets are never too perfect. This approach works with a complex field like viral immunity, but it isn't appropriate, of course, for a mechanistic analysis like determining the sequence of a gene or a protein. The sequence has to be correct. The analysis is readily repeatable by anyone with the right expertise and equipment and there are real penalties in terms of the individual's reputation for being wrong.

However, if the results are informative and the study repeats, we might then go on to ask a further question, or may decide that it looks pretty good as it is. The next step is then to write the scientific paper and report the results for the scrutiny of our colleagues and competitors, our peers (the process termed 'peer review'). As Karl Popper, the influential philosopher of science, pointed out, only the written record is valid in science. No matter how honest we may be, memory is easily distorted. Besides, some of the key insights may not emerge until we look really closely at the research data and assemble the findings into a convincing story.

It is important to discuss ahead of time how any particular study will operate in terms of who pulls all the data together and does the initial write-up. That person will normally have her or his name as the first author on the eventual research paper, a point that needs agreement from everyone involved. Order of authorship is enormously important in the business of credit for biomedical scientists. In biology, at least, the usual convention is that the first name on the manuscript is that of the junior scientist who has the greatest responsibility for, and involvement in, the particular project, while the PI is the last, or senior, author. Sometimes the technical work of the study divides equally between two young scientists, so the agreement is that they will alternate as first author on successive research reports. Running a successful research laboratory demands both time and attention. Because of my other commitments, I now work directly with experienced colleagues, and will often put my name in the middle rather than at the end of a list of authors. My program at the University of Melbourne is operated jointly with Steve Turner, an Australian who worked with me as a postdoc at St Jude, then returned to establish the laboratory and he is now an independent university faculty member.

I am sometimes asked why I continue to operate two such physically distant research efforts. The great strength of the St Jude program is the continued collaboration with the virologists Rob Webster and Richard Webby, which allows us to work with viruses that have been engineered to address unique questions concerning the influenza-specific immune response. Rob also runs one of the few bio-security laboratories where it is possible to do protection experiments with the extraordinarily lethal avian influenza viruses that are a looming threat to humanity (there is more on this in the next chapter). Melbourne, on the other hand, offers a sophisticated immunology community where we have developed a spectrum of effective collaborations with structural biologists and molecular geneticists interested in the regulation of immune effector mechanisms and the nature of T cell memory (see chapters 4 and 8). In fact, we exchange both reagents and personnel between the two locations. Cutting-edge science is an international activity that is driven, in part, by the principle of selective advantage. So, though the arrangement may perhaps seem cumbersome, it works, the science benefits and a few very talented young Australians and Americans gain in their level of international experience and exposure.

Early in my career I wrote pretty much every word of anything I ever published. Nowadays my starting point is often a draft manuscript from a young associate. As I agonise over and re-work the text, I am constantly checking back with questions like 'How exactly was this done?', 'Do we have a repeat experiment?', 'What do you think of this result?', 'Could it mean that …?', 'Do you remember who said …?' We eventually end up with something that is mutually acceptable and, hopefully, understandable by others in the field. The next step will be to submit our manuscript of 1,000–2,000 words of carefully reasoned argument, plus diligently cited references to other studies, and illustrative figures and tables full of results. Sometimes we may go for a rapid publication format, but if we aren't in too much of a hurry, our latest 'baby' generally will go to the most prestigious scientific or professional journal that we think might accept it.

Sometimes we write longer, solicited reviews of our own work or of the field in general; these are usually under 5,000 words long and rarely more than 10,000. Most research journals appear monthly. This can also be the case for short review formats, though the longer accounts come out as annuals. Timely, specialist reviews are very important, as the magnitude of the scientific literature is such that only those intimately involved in a particular sub-field are likely to have either the time or the inclination to read the primary research reports.

A submitted paper will often bounce straight back with 'thanks, but no thanks' from the editors of one of the top journals, like the weeklies *Nature* or *Science* that are as much newspapers as research report formats. If the editors decide after a preliminary read that they might want to publish, the article will then be sent to two or three informed colleagues for anonymous peer review. They may have lethal criticisms that will ensure rejection or, quite commonly, they may suggest additional experiments that need to be done or arguments that should be made. Anonymous review can be open to abuse, but most behave responsibly as everyone lives by the same rules. In my experience, the famous aren't cut any slack in the peer review process, and indeed they shouldn't be. The integrity rests in the validity of the data and the intrinsic interest of the conclusions, not in the personality or prominence of a particular author.

Publication means credibility when it comes to securing research grants, jobs and promotions. Science is very competitive and there is always the possibility of being beaten to the post. Credit for a big discovery that can lead to a Nobel Prize requires the recipient to be first to publish the primary research data or the spectacular new idea, even though the report of the actual discovery can be very brief indeed. The 1953 paper on the DNA double helix by Jim Watson and Francis Crick that started the molecular biology revolution occupied fewer than two pages of *Nature*. The key findings and ideas that led to our 1996

Medicine Prize appeared in three research reports (reproduced here in the appendices) that took up fewer than four pages in the same journal, and in a four-page hypothesis article that appeared in *The Lancet*.

After we left Canberra in 1975 for separate careers in the United States, Rolf Zinkernagel and I unwittingly devised and conducted sets of experiments that were similar, reached the same conclusions and sent them off for review. Mine was accepted by *Nature*, his by the *Proceedings of the National Academy of Sciences*, which are both classy journals. By this time, we had talked and knew what had happened. His paper came out and I waited, and waited, and waited. It transpired that my manuscript had been buried on someone's desk at *Nature* and it didn't appear until the following year. The protection in the priority game is, however, that every published paper also carries the date of acceptance by the journal. If someone was comparing two candidates for an award very closely, this would be one way to sort it out. Has anyone ever missed out on a Nobel Prize because of something like this? I doubt it, though it would be an interesting question for a science historian to look at now that the archives of fifty years of the committee proceedings are open for review.

What experienced scientists do when it comes to the stage of writing-up is to start by laying out the experimental results in tables and figures, then develop a closely reasoned account of the data set that provides the basis of the story that is to be told. The account may well support the idea that stimulated the study in the first place, but it can equally be true that a novel conceptual thread emerges as a consequence of looking at the new evidence that has been generated. This is then reflected in a discussion of the results and in the summarising abstract, which should convey the strongest points in an exciting way. The last step may well be the introduction, laying out why this particular study was done. At this point a research paper can be a little disingenuous: the introduction may reflect more what we might have thought if we had been really smart and knew what the results were going to be, rather than the ideas we had when we set out on the journey.

We can find ourselves going down the road of what philosophers describe as 'the Texas sharpshooter fallacy': Tex stands back, takes aim and empties his six-gun into the wall of a barn. He then draws the target around the clustered holes in the barn wall. This is a great way to score a bullseye. If our first experiment ends up defining a shot pattern that's a long way off our original aim, we will only confuse readers if we introduce the research paper with our reasons for shooting at the other target in the first place. A bullseye that looks a lot better than anything we might have hoped for can lead to our subsequent experiments taking a very different direction. Tex's bullseye may, in fact, be serendipitous—and a novel discovery may have been made by chance.

The 'firearm–barn wall' strategy has recently been formalised by the description 'discovery science'. Generally, this has been defined as 'science that is not limited by an hypothesis'—and the processes will usually lead to a mass of new questions which will then need to be followed up by the usual, hypothesis-driven research strategies. If we think of firing an unchoked 12-gauge shotgun loaded with birdshot into our barn wall, the resultant splatter of pellet holes is the 'discovery profile'. It may be all over the place, but there is a pattern and the job may be to work out what that means. The hole made by any given pellet may look different from what might be expected, and could represent the beginning of a new 'reductionist' science project that advances by sequential hypothesis and experiment—or even the theme of a future research career.

Discovery science is coming into its own in biology, and particularly in the new science of genomics. A century after the first Nobel Prize, the efforts of people like Craig Venter, Francis Collins and others resulted in the first published sequence of the human genome, the complete DNA code that ultimately determines the nature of our physical beings. The mouse genome soon followed. There are now published genomes of many life forms, ranging from the malaria parasite to the fruit fly to the chimpanzee, the chimpanzee's being about 98 per cent identical to ours. Knowing the genome can perhaps be likened to having a telephone directory that lacks many of the names and addresses. The massive task ahead of us is to associate the numbers with names, then work out what these individuals do and how they fit together.

*B*oth the PhD and the postdoc periods of a scientist's life are intended to be training experiences. Part of the training is to become an independent scientist. Another important step is to learn how to work with others and to make the best use of whatever technical assistance may be available. The degree of direct involvement by the PI is likely to reflect the maturity and independence of the trainee. The cultures within 'small science' laboratories are as different as the personalities and characters of the key players. Some PIs are quite controlling and will want to supervise the planning of every experiment very closely. They may also require each young scientist to work on a very different project and report only to them. Others are more open and hands-off.

My personal style is to encourage people to collaborate. Viral immunity is particularly complex and any one experiment can require the 'person-power' to look simultaneously at a number of diverse variables measured by different techniques. Though each young scientist will do individual small, often preliminary, experiments that probe their own specific ideas and refine their technical abilities, the capabilities they develop also contribute to larger investigations. A big experiment may typically start at seven in the morning and, because there are a number of time-consuming steps that require incubations, staining and so forth, the day's work may not be completed until midnight. People make mistakes when they are exhausted, so groups often arrange to operate in different shifts.

The lead player in any particular experiment may be in a support role for another investigation. My job as a PI might be to bring the coffee and doughnuts, and to discuss any last-minute changes in the experimental protocol that result from some unforeseen difficulty. Though I served my apprenticeship and worked for years at the bench, any 22-year-old technician will do hands-on laboratory work much more quickly and efficiently than I can. Having me directly involved in an experiment would be the equivalent of asking Dr Manette in *A Tale of Two Cities* to run a competitive shoe-repair business in his somewhat decrepit years after being released from the Bastille. Manette was a real doctor, who had to become a boot-maker during his eighteen years of unjust imprisonment.

The initial success that every scientist must achieve is to make the transition from postdoc status to establishing their own independent laboratory. Many institutions, particularly in Europe, will try to retain extraordinarily talented people who emerge from within strong research programs. However, the usual situation in the United States is that people move on at this stage of their career. There is a constant search for bright young candidates—and talent is what matters. Some young scientists become fixated on going only to a high-prestige institution. My approach in this has always been a bit idiosyncratic in that I have been happiest in situations that had a strong collegial feel, where I also had the sense that I would be able to concentrate on my research. Others prefer to work in high-status, very competitive environments.

With e-mail, and with FedEx to transfer samples for analysis, it is easy to have strong collaborations at a distance, at least within the United States or Europe. Willy Allan, a Scots postdoc in my Memphis laboratory, conducted a whole series of experiments on influenza pneumonia with Simon Carding (from Yorkshire) who was then a postdoc in Kim Bottomly's immunology laboratory at Yale University in New Haven, Connecticut. The mouse experiments were done at St Jude in Memphis, then the slides were couriered to Simon, who processed them by a technique called *in situ* hybridisation to determine the read-outs for particular lymphokine genes (see chapter 4). A number of research papers were published, Simon's career was helped along a bit, and he is now back in Yorkshire as a professor at Leeds University in the United Kingdom working on inflammatory bowel disorder. Willy decided that life in a research laboratory was not for him and is now a successful podiatrist in Perth, Western Australia. Moving to podiatry is certainly a bit unusual for someone like this, but jobs for life are a thing of the past and many people now change careers in mid-stream.

As young scientists come to the end of their post-doctoral training, some decide that though they might wish to continue in research, they would be more comfortable in an industry setting. The latter is likely to be better paid and, though it was almost unknown twenty years ago, it isn't uncommon now for biomedical PhD candidates to start out with an industry career in mind. This has long been the case in engineering, and institutions like MIT have a formidable record of producing graduates who turn scientific discoveries into economic opportunities. Many who choose to go down the industry road move fairly quickly into management or even sales, and end up living more of a business life. At the University of Melbourne, for instance, Dick Wettenhall has established an undergraduate biotechnology course that orients school leavers in this direction from the beginning. Joint business–science or economics–science degree courses are increasingly popular as the realisation dawns that high technology will drive a great deal of future industrial activity.

The other essential for the emerging scientist is to become established within a particular scientific discipline, then to be prepared to go on the circuit, give seminars, participate in scientific meetings, present talks and ask questions. Most will have had the experience of attending major international conferences annually since their student days, meeting with established investigators and giving presentations. Like any other human activity, it helps enormously if a name on a technical report or review becomes attached to a face and a personality in the minds of the key players in a field. So you've got to be multi-skilled: the work of the experimental scientist covers the spectrum from student to public speaker, from funding generator to administrator of all aspects of very substantial programs. Then there is service on various national and international committees, though it's important for anyone who is working creatively to avoid being consumed by committee activities.

Day to day, the scientific life is exciting and fulfilling. While most scientists live ordinary lives with spouses, children and mortgages, they are likely to work irregular hours, particularly when they are more junior. All scientists tend to determine their own working day, and most laboratory heads have no problem with graduate students or postdocs who, providing they attend laboratory meetings and fit in with the demands of group experiments, operate on their own schedule. Big experiments often benefit from the involvement of people who like to start and finish at different times of the day. Some senior investigators may come in at midday and work late into the night. The international cooperation and competition that are an essential part of science also mean frequent communication with those pursuing similar research questions in other countries. Sometimes a chance meeting at a symposium in the Rockies or a meeting on an Adriatic island will produce a long-distance collaboration with studies done in one place and samples evaluated in another; a five-minute conversation can also mean the difference between moving down the right path or spending months going in a wrong direction.

Smart people who are working on similar problems will often come to the same conclusion. I've sometimes decided not to take a particular course when I learned from a brief chat or a formal talk that what I had been thinking about was already well underway elsewhere. Another great clarifying experience, and increasingly a feature of the scientific meeting circuit, is seeing some of the extraordinary presentations made by people like Mark Davis, from Stanford, or Ulrich von Andrian, from Harvard, who use contemporary imaging approaches to provide visual insights into, for example, the ways that the cells involved in immune responses behave (see chapter 4). If one picture is worth a thousand words, one short movie can be worth a thousand pictures in a research journal. Such material is increasingly available from the Web, but the first encounter often comes from the totally unexpected experience of seeing a sixty-second clip in a scientific talk.

At the beginning of a scientific career, of course, the money is not high, but stipends for bright young PhD students are generally paid at a level that is sufficient to maintain life. Support may come from a national scheme, from training grants to substantial universities or from the budget of a research grant to a senior scientist. Graduate students often supplement their income by undergraduate teaching or tutoring. Postdoctoral fellows are paid considerably more. After they're established, most scientists live reasonably well, and job security is generally good for those who achieve long-term university positions. Maintaining both tenure and a high salary in a top research university will inevitably mean being competitive in the research grant scene, though some leave the laboratory and move on into university administration. Scientists are increasingly prominent as university presidents or vice-chancellors, usually the best-paid jobs in any academic setting.

A few scientists become very wealthy. All academic institutions have mechanisms that ensure that anyone who makes a commercially important discovery will at least share in the resulting profits. Some launch their own companies. Even if they didn't achieve fame and fortune, most scientists in the latter part of the twentieth century benefited from the first mass 'culture of creativity' in the history of humanity, as governments facilitated and funded discovery and the generation of knowledge. The success of the enterprise is, perhaps, reflected in the fact that elements in the entrepreneurial-business culture now focus heavily on biology and biotechnology. Though it was not the case for researchers of my generation, who have had to learn such matters late in life, young scientists are increasingly well educated on issues relating to intellectual property rights, patents and so forth. In turn, young lawyers and business trainees are being drawn more into science-related areas. There were in the past, as described in Dava Sobel's *Longitude*, occasionally nationally resourced competitions to seek solutions for important national problems, but government-funded research on a broad scale is a recent phenomenon.

Scientific discoveries emerge when the research investigator can essentially 'play' with ideas and experiments. Politicians aren't necessarily enthusiastic about the idea of spending tax dollars so that smart people can play, even if the 'play' goes on in the context of trying to understand, and deal with, problems like AIDS and cancer. Like it or not, that's the process that leads ultimately to new technologies and therapies. Once the political process tries to direct research, disaster inevitably results. Scientists at the top are no longer funded, and the politicians end up paying third-rate 'cannon-builders to put a man on Mars'. It takes a sophisticated political process to deal with this reality. The US style of government, where the division of powers between the executive (President and Cabinet) and the legislative branch (House and Senate) makes for a quite open political dialogue, may be the ideal form of government for promoting innovation. Parliamentary systems like Australia's are much more rigid and susceptible to dictatorial fiats and collective, ideology-driven fantasies when it comes to funding science.

Most senior investigators are bright and effective people with a record of consistent, high-level achievement. The populations of Nobel laureates and leading research academics overlap, of course: the difference is that the Nobel Prizes go to scientists who have had the good luck to hit on a stellar, breakthrough discovery, or the intellect to make some immensely powerful theoretical contribution. Most will be long-term achievers, but this is not always the case. Nobel Prizes aren't won by simply accumulating scientific brownie points. The prospect of winning prizes isn't why people go into science and stay there. When they reach 65, many successful scientists are still so excited by what they do that they choose not to quit. There are plenty in their seventies who are still seriously committed to research and the promotion of science—while others walk away from the job like people elsewhere, and find new interests. Fred Sanger, who won two Nobel Prizes, quit at 65 to grow roses and sail his yacht. He'd done what he came to do.

So, science isn't about achieving high-sounding positions, or prizes, or riches. It's about discovery and excitement. It's also about persistence and integrity. Perhaps the best advice that can be given to any young scientist is: commune with the data. Look at the findings over and over from different perspectives: if what you have seems odd but interesting, try lateral thinking: think the impossible or the absurd. Talk with trusted colleagues. The mark of the creative scientist can often be to see something new in findings that others have dismissed as uninteresting, or a failure. Sometimes the hardest thing for any of us to see is what is directly in front of our own faces, particularly if we are locked in conceptually to a particular theory or set of ideas.

On the other hand, first-class scientists know how to avoid going down unproductive paths. I have met more than one researcher who persistently followed the abnormal result and, just as consistently, picked the path to the red herring and certain failure. If you do find something unusual, it is important to see that the result can be repeated before investing too much energy. Finding out new stuff is a demanding game, and what looks like serendipity can sometimes turn out to be the song of the sirens from Greek mythology.

The greatest concern for anyone heading a research program is that somebody will cheat and fake the result—though it hasn't, as far as I know, happened to me personally. It's rare, but it can occur when someone young and inexperienced somehow comes to believe that the job of the person doing the experiment is to support the ideas and interpretations of the leader. Nothing could be further from the truth. My young colleagues are made very aware that their task is as much to deflate any grand constructs and pretensions that I might have as to reinforce them.

I am delighted when they are able to convince me my idea is wrong. Part of my task as a senior scientist is to help them to emerge as the next generation of innovative thinkers and investigators. They have to grow and to become independent.

Scientists do make mistakes and the less scrupulous can and do lie when it comes to the business of allocating credit, particularly for the source of an idea. However, the actual scientific data cannot be faked. Apart from being a matter for criminal investigation if the research is funded by federal research grants, for instance, developing a further set of experiments or theoretical interpretation on the basis of wrong information will inevitably lead to disaster. A long-term scientific investigation can be thought of as an ever-branching tree. If the trunk is rotten, the whole tree will fall. Years, lives and millions of dollars can be wasted. The worst scenario is when evidence emerges that primary research data has been manipulated, or even invented. In the United States this can lead ultimately to an FBI investigation, but the truth must out no matter how painful the consequences.

The great molecular biologist David Baltimore, who shared the 1975 Nobel Medicine Prize, became involved in just such an affair which began in 1986 when a junior colleague, Tereza Imanishi-Kari, was accused by a 'whistle blower' of publishing fraudulent data. David's defence of Tereza contributed to his leaving the presidency of the Rockefeller University (New York), though for some years he has been heading Caltech, in Pasadena, California, where he would certainly enjoy better winters. Tereza was eventually exonerated, but not before she had lost ten years of her life as a productive scientist. The story is told in

Shane Crotty's *Ahead of the Curve: David Baltimore's Life in Science* (University of California Press).

Sometimes, however, there is indeed clear and irrefutable evidence of deception. A famous instance is the case of the painted mice. The story is told in Joseph Hixson's *A Patchwork Mouse* (Anchor Press) and by Sir Peter Medawar in a chapter entitled 'The Strange Case of the Spotted Mouse' in his book *The Threat and the Glory* (HarperCollins). A young medical scientist generated some exciting evidence of long-term survival when he first cultured black mouse skin tissue in a particular way, then grafted it across a histocompatibility barrier onto a white mouse. Normally, of course, the 'host' white mouse would have rapidly rejected the 'donor' black skin (see chapter 4). After a week or two, only an all-but-undetectable scar should remain. As others were to show later, his first, promising results may well have been right. The problem was that he could not repeat them and eventually faked the result in a quite crude way by painting a 'patch' on a white mouse with a black marker pen. Not surprisingly, he was soon found out. Many cases of fraud show this same profile of a 'good' early result, followed by a failure to repeat, although there are also a few cases where there was deliberate deception from the outset. Sociopaths are found in all walks of life, and science is no exception.

Science fraud generally occurs in high-profile institutions, often in the medical research area where expectations can sometimes be raised way beyond the scope of the available evidence. There have been prominent instances in both the United States and Europe, and I know of one such story in Australia. I was peripherally involved in this case because, while I was living and working in Philadelphia, I was asked to review the research report for publication as a letter to *Nature*. I knew the PI to be a first-class scientist, and both he and I were fooled by the exciting 'data' that his young colleague had generated. The article was published, though the deception was soon discovered and the claims were immediately withdrawn.

Of course, most mistakes in science don't come from deliberate fraud. This is one reason we repeat experiments. An interpretation can be wrong because the technology was imperfect and unknown variables were at play. The mistake will become apparent when new approaches or instruments become available. When it is recognised that an idea based on a flawed finding has gained some credibility in a field, it is essential that the originator make the effort to point out very publicly why and where the hypothesis was wrong. This is as much a matter of self-protection as anything else. If the initial protagonist fails to do this, there is an absolute certainty that someone else will.

Joint programs can be an enormously effective way to do science. The Nobel Prizes for Medicine to Mike Bishop and Harold Varmus in 1989 'for their discovery of the cellular origin of retroviral oncogenes', the genetic elements that contribute to the development of cancer, and to Joe Goldstein and Michael Brown in 1985 for their work on cholesterol metabolism, reflect efforts of this type. Though Rolf Zinkernagel and I shared the 1996 award, we worked together for only about two-and-a-half years at the beginning of our respective careers in immunology.

Programs led jointly by a wife and husband can also work well with, not uncommonly, the woman scientist taking the more public profile and the man staying closer to the laboratory bench. The most famous Nobel couple is, of course, the Paris team of Pierre Curie and Marie Sklodowska Curie, who shared the 1903 Physics Prize for radioactivity with Henri Becquerel. Their daughter, Irene Joliot Curie was, in turn, awarded the 1935 Chemistry Prize with her husband, Frederic Joliot, for the discovery of the radioisotopes that remain so important in the research laboratory, diagnostic medicine and therapeutics. The expatriate Czech scientists Carl and Gerti Cory shared half the 1947 Medicine award for their discoveries concerning energy metabolism, while working at Washington University of St Louis, in Missouri.

Perhaps the best known husband and wife team in my own field of immunology are Philippa (Pippa) Marrack and John Kappler; they are both HHMI investigators (see chapter 6) working at the National Jewish Medical and Research Center in Denver, Colorado. Pippa is English and Cambridge-educated, and John is American. The *New York Times* science journalist Gina Kolata called Pippa the day after the announcement of our Nobel award to find out who these Zinkernagel and Doherty characters were. Her reply was that Zinkernagel was fairly normal but that Doherty was a bit 'Eeyore-like'. For anyone who doesn't know, Eeyore is the likeable but depressive donkey in A. A. Milne's *Winnie the Pooh*. The net result is that I now have one of the world's most comprehensive Eeyore collections, ranging from key rings, through a *Little Golden Book* to two large stuffed blue versions that differ because one talks when its leg is squeezed. Our little granddaughter, Julia, loves the talking Eeyore, which was given to me after a lecture at the University of Southern Queensland, and it's the first thing she checks out when she visits.

A few months after the award, I was interviewed in Memphis by 11-year-old Arundathi Gururrajan for her school newspaper. Her interview also appeared later in India's national daily, *The Hindu*. Arundathi was very sweet and came to the conclusion that I was more A. A. Milne's bouncy Tigger type, so I now have one, though only one, toy plastic tiger, a gift from Arundathi. I expect that, like most people, I'm part-Tigger, part-Eeyore.

Early Influences

What sort of family experience might be expected to form a future Nobel Prize winner? The pattern is that there's no pattern. There are a few Nobel families. Apart from the father, mother, daughter combination of Pierre, Marie and Irene Curie, the father and son team of William and Lawrence Bragg was awarded the 1915 Physics Prize, and the chemist and neuroscientist Ulf von Euler (Medicine, 1970) followed his father, the enzyme chemist Hans von Euler-Chelpin (Chemistry, 1929). Fritz Zernik (Physics, 1953), who invented the phase contrast microscope which I used in my early research, was the uncle of Gerardus 't Hooft (Physics, 1999). Other Nobel laureates in science were also raised in academic families, and some came from prosperous business backgrounds. Many grew up in very modest circumstances and were the first in their family to attend a university, or even to complete high school. Science is, after all, a common 'up and out' path for bright kids who come from a social context where there is little understanding of the realities of power and privilege.

The Nobel laureates in science also went to all sorts of schools. Some spent their childhood on farms where the quality of the available, formal education was limited. A few were home-schooled. Because most are from the United States or Europe, the majority attended taxpayer-funded primary and secondary schools, or Catholic schools of equivalent wealth and status. Richard Axel (Medicine, 2004) is a graduate of Stuyvesant High School, joining a list of twenty-four others from the New York City public school system, five of them from the Bronx Science High School. This fantastic record probably reflects that many of the Jewish refugees from Germany and eastern Europe did not move beyond New York City.

Britain and Australia both have school systems that are divided into expensive private institutions (called 'public' schools for historical reasons), some other less costly religious schools and also state-funded schools of varying stature. A few British Nobel laureates went to the great public schools, including Peter Medawar (Medicine, 1960) and John Sulston (Medicine 2002). Many others were educated in grammar schools after passing the competitive 11-plus examination, a model of selective schooling that has now been consigned to history. Of the six science laureates who had their schooling in Australia, both Lawrence Bragg (Chemistry, 1915) and Howard Florey (Medicine, 1945) attended Adelaide's St Peter's Collegiate School, and Mac Burnet was at the prestigious Geelong College. John Cornforth and Jack Eccles were at Sydney Boys' High and Melbourne High respectively, which are selective, single-sex state schools, and I went to a state high school that took any boy or girl from Brisbane's western suburbs.

It makes sense that a high-quality selective schooling, whether the selection is based on the merit of the child or the economic circumstances of parents, does foster talent, but it is also apparent from the record (www.Nobelprizes.org) that this is not an essential pre-requisite for later academic stardom. Many of the Nobel biographies mention an inspiring science teacher. In John Cornforth's case it was Leonard Basser; Paul Nurse (Medicine, 2001) names Keith Neal, and Tim Hunt (Medicine, 2001) Terence Doherty, who is no relation. Lawrence Bragg and Howard Florey had the same chemistry teacher, 'Sneaker' Thompson, so-called because he wore rubber-soled shoes in an era when that was uncommon. We can never value dedicated teachers too highly. There are also other influences of childhood that a number recall, like learning music, playing with creative toys such as Meccano sets (known as Erector in the United States), or finding a passion for nature and collecting life forms, such as beetles in Burnet's case.

I also looked at the biographies of the science laureates over the past twenty years to see if there was any determining pattern in the colleges and universities that they attended. Again, being accepted at a top institution is likely to be a plus, but it isn't essential. While 40 per cent of the British laureates were at Oxford and Cambridge, another 40 per cent were at Birmingham, Manchester or Sheffield. Oxbridge was, of course, a bit slow off the mark in developing strong science programs during the nineteenth century, while the newer 'red bricks', like Birmingham and Manchester, went in that direction from the outset.

The United States is unique because, in addition to great state universities like UCLA, Berkeley, the University of

Washington, the University of Michigan and the University of North Carolina, it has a number of first-class private tertiary institutions. These range from the north-eastern Ivy League universities, such as Harvard, Yale, Cornell and Penn (the University of Pennsylvania), to equivalent campuses in middle America (the University of Chicago, Washington University of St Louis, Vanderbilt, Emory) or on the west coast (Stanford, Caltech), to the small (fewer than 2,000 students) liberal arts colleges (Haverford, Pomona, Swarthmore, Rhodes) that emphasise a high-quality teaching experience modelled on the Oxbridge college system. The biggest sample population in my twenty-year Nobel list was, of course, the US set, and of these laureates, about a third went to state universities and two-thirds to private institutions. Of the latter, a third were at liberal arts colleges. Though the top private universities charge high fees, they also offer a lot of scholarships for those with exceptional talent, whether in academics or on the sports field, so this distribution pattern is by no means an indication that the leading US scientists come from wealthy families. Some do, most don't.

Scientists have to learn an enormous body of highly specialised knowledge before they can pursue their craft. The university and college systems I've mentioned above provide the training for the first step in a science career, which is generally the Bachelor of Science (BSc or BS) degree achieved after four years of undergraduate study. The next step is to gain the 'science ticket', the Doctor of Philosophy (PhD or DPhil) degree that requires the candidate to do research under the broad supervision of an established scientist. Again, this will generally go on in a major university, though there are also 'partnering' arrangements that allow students to work in top research institutes like the laboratories of the Institut Pasteur, the Max Planck Society, or places like the Walter and Eliza Hall for Medical Research, Melbourne, and St Jude Children's Research Hospital. The hands-on apprenticeship of the PhD requires the candidate to work 'at the bench', do original research, publish in the peer-reviewed scientific literature and write a comprehensive thesis.

The bench for the theoretician might be the blackboard and the computer; the physicist may design at least a component of the experimental equipment, whereas the chemist and the biologist are likely to be involved with the flasks, tubes, pipettes and the lab coats that we can all recognise from television and the movies. Those who like to work with their hands often have the time of their life during the PhD years. The administration and competition for grants that also goes with a scientific career is still in the future. At this stage, it is a big plus to be in a dynamic, first-class institution where there are many bright people around to learn from and talk to.

One difference between the United States and the rest of the world is that the undergraduate experience is dedicated to the idea of providing a liberal education, which means that those working towards a BSc degree are also required to take English, a foreign language, history and so forth. Even future literary writers have to work their way through courses with names like 'Physics for Poets'. I am very much in favour of this approach, as it does allow young people to grow up a bit before being locked in to a long-term career. However, it also means that there is less time for in-depth science teaching, and some of the defects have to be made up by further course work when it comes to the PhD program. While a typical Australian or British PhD may take three or four years, an American doctorate often requires five or six.

The formal lectures and tutorials given in a good American PhD program can be of great benefit for those who come from less well-resourced universities and nations. Many top US scientists have had this immigrant experience. The spectroscopist Ahmed Zewail went from the University of Alexandria in Egypt to Penn for his PhD, and was at CalTech when awarded the 1999 Chemistry Prize. Mario Molina, who shared the 1995 Chemistry Prize with Paul Crutzen and Sherwood Rowland for the CFC-ozone-layer depletion story, graduated from the National University of Mexico, then took his PhD from Berkeley. Even in a sophisticated culture like Japan of the 1960s, Susumu Tonegawa (Medicine, 1987) was advised to go from Kyoto University to the United States for his PhD in molecular biology at the University of California, San Diego.

Not every future Nobel laureate followed a typical academic path, and some didn't even complete a research PhD. This is most common for medical graduates, where the MD plus a couple of years in a first-class laboratory is often enough to set people up for a stellar research career. Of the medical Nobel laureates reviewed in my twenty-year survey, 60 per cent had a PhD alone, 30 per cent an MD alone, and the remainder had both degrees. Of the US laureates, some 40 per cent list only an MD.

A few of the physical scientists, particularly those with an engineering background who went straight into industry, did not bother with any form of doctoral degree, though all will have been awarded honorary doctorates by universities wishing to hear from them at undergraduate graduation ceremonies. The much-loved pharmacologist

Gertrude (Gertie) Elion took a masters degree from New York University, then decided to stay in a good position at the Wellcome Research Laboratories rather than go back to a university. She was 70 when she shared the 1988 Medicine Prize with her North Carolina colleague George Hitchings and the Scots scientist Sir James Black.

As with any human activity, there are exceptions that prove the rule. It generally makes sense for the trainee scientist to work in the best available places with the smartest people. Sometimes, though, the way to go may be as Robert Frost suggested:

> Two roads diverged in a wood, and I—
> I took the one less traveled by,
> And that has made all the difference.

There are no absolute standards that determine where, how and by whom creative science is done.

4

Immunity: A Science Story

A little like the song lines running through Indigenous culture, several distinct themes can be traced in the hundred-year history of the Nobel Prizes for science. A difference is, though, that the end story in science is constantly being rewritten, and what goes before is reinterpreted accordingly. What follows is a brief account of the Nobel lineage in one scientific discipline, my own field of immunology. Understanding this story should give you some insight into how biological science, in particular, developed through the twentieth century. Telling it also allows me to put in perspective the award that I shared with Rolf Zinkernagel. Anyone who might wish to find out more about the science and the scientists discussed here can, of course, go to the Nobel website (www.Nobelprize.org) and read the presentation speeches, the brief autobiographies and the Nobel lectures delivered by the recipients.

The two great complex biological systems that we use to deal with the infinitely diverse world that surrounds us are the central nervous system (CNS), the brain and its attached sensors like the eyes and ears, and the immune system, which operates, of course, below the level of consciousness. We all have some familiarity with brain function, but immunity is much more mysterious and obscure.

Before I tell the story of immunology and the Nobel Prizes, it is first necessary give a short account of how the immune system works. I will do my best not to use the specialised language of 'immunology-speak', though I fear it can't be totally avoided. Many of the examples are drawn from viral immunity, the subject of my own research.

The word 'immunology' is derived from the Latin *immunis*, which described the status of some returned soldiers in the Roman Empire who were, for a time, exempt from paying tax. The tax that the immune system functions to alleviate is that caused by parasitism, the growth of much simpler organisms like viruses, bacteria, fungi or worms in or on the more complex higher vertebrates like us. In the absence of effective immune control we may die from poisoning with bacterial toxins, from having our cells and organs destroyed by viruses or as a result of being overwhelmed by the uncontrolled growth of alien invaders. This is what can happen to people who have AIDS, or who are severely immunosuppressed by the radiation treatment and chemotherapy that is used to kill cancer cells. As Susan Sontag points out in *Illness as Metaphor*, we use the language of war and aggression when we talk about disease and, at least in the case of infection and immunity, the analogy is not a bad one.

The first line of defence when an infectious agent enters our body is mounted by what we call the innate immune system. Innate immunity goes way back in evolutionary time. Human beings share the protection mechanisms that operate in fruit flies and even in single-cell life forms like amoebae. The only Nobel Prize awarded in this area went to Ilya Mechnikov in 1908 for his discovery of phagocytosis. The phagocytes, or macrophages (big eaters), engulf and then destroy micro-organisms or our own damaged cells. A black skin tattoo, for example, may consist largely of carbon particles sitting around in long-lived macrophages. The macrophages 'eat' the injected carbon granules dispersed in the tattoo ink, then park themselves in the skin where they survive indefinitely.

The precursors of the macrophages are the monocytes, a subset of the white blood cells that are made continuously in the bone marrow. In one way or another the monocyte/macrophages are involved in just about every immune response, even if it's only as scavengers to clean up after the destruction that is the inevitable consequence of the elimination of unwanted cells and invading organisms. A pattern of general involvement is also true for other mediators of innate immunity, like the 'natural killer' cells and the neutrophils, though I won't discuss their roles in this brief account. The field of innate immunity is currently advancing very rapidly and has a good chance of being recognised by at least one further Nobel award.

Most of us are more familiar with what specialists call adaptive immunity, the specific and robust host defence system that is found only in the higher vertebrates. The evolutionary biologists, in fact, date this adaptive response back to the jawed fishes which first emerged about 350 to 400 million years ago. That may seem a long time when we consider that our species, *Homo sapiens*, has been around for only 150,000 to 200,000 years, but it's a fairly short interval in the long march of biology. The brain, for instance, goes back considerably further in evolutionary time than the adaptive immune response.

When we think about immunity, the minds of many will turn to the childhood experience of being stuck with a needle to deliver this or that vaccine. Vaccinations protect for a long time and are specific for the particular infectious agent, whether it be a virus, worm or fungus. The first documented vaccine was the "cowpox" that Edward Jenner took from the hand of a milkmaid, Sarah Nelmes, who had been accidentally infected from the teat of a cow in 1796, and scarified into the arm of young James Phipps as a means of preventing smallpox. Jenner then challenged James with fully virulent smallpox virus, an experiment that would obviously not be allowed today. Fortunately James survived and, nearly 200 years later, after a massive international campaign using approaches that were not so different from Jenner's, the World Health Organization (WHO) of the United Nations declared in 1980 that smallpox had been eradicated globally.

The eminent virologist Frank Fenner was recognised by both the Australian Prime Minister's Prize and the Japan Prize for the part that he played, as chair of the responsible WHO committee, in the smallpox eradication campaign. Frank is greatly loved and respected, and we were delighted that (then in his eighties) he was able to come to Stockholm to represent the ANU on the occasion of our 1996 Nobel award. Despite the fact that smallpox is the only disease ever known to have been eliminated from the planet, there is still great concern at the appalling possibility that bio-terrorists could resurrect either the smallpox virus or a more lethal variant that has been genetically engineered in some way. As a consequence, efforts to develop an even more effective smallpox vaccine are continuing.

It was hoped that polio (poliomyelitis) would also be history by now, but immunisation programs in Africa were disrupted by local wars and, for a time, by resistance from an isolated group of religious leaders who were manipulated politically into believing that the virus had been engineered to prevent female fertility. That's nonsense of course but, like the GM food story that I discuss in another chapter, the topic of vaccination is surrounded with more misconceptions, conspiracy theories, prejudice and misunderstanding than almost any other area of medicine. Look up 'vaccines' on Google if you want to get an idea of this. Part of the difficulty is that we administer these products to very young children, who don't much enjoy the experience and may feel poorly for a day or two afterwards. Another is that millions of children are vaccinated each year. Given the numbers, it's inevitable that a few will develop symptoms of something or other over the following weeks. The tendency then is to blame the vaccine. The irony is that vaccines, along with the introduction of sanitation and basic hygiene, are the cheapest and most effective medical interventions ever developed. There is, though, as with any medical intervention, a risk–benefit ratio, with the benefit in favour of vaccination being enormous. Incidentally, so that I don't confuse anybody, the terms 'immunisation' and 'vaccination' are used interchangeably in medicine.

The Salk polio vaccine is made from fully infectious poliovirus that is first grown in tissue culture, killed (inactivated) by exposure to formalin, then clarified and made non-toxic for injection into young children. The Sabin vaccine, on the other hand, is an attenuated, or 'weakened', strain of the live virus, which still multiplies to a limited extent in the cells of the upper gastrointestinal tract. The dead Salk vaccine must be injected, whereas the live Sabin virus can be given by mouth on a lump of sugar. For this reason the Sabin vaccine can be dangerous for those rare infants who have no effective immune system as a result of some genetically determined immunodeficiency disease, like the so-called 'baby in the bubble'. Even the 'mild' Sabin virus will continue to grow and, in the total absence of adaptive immunity, cause disease.

Poliomyelitis vaccine works by stimulating the production of specific antibodies that bind to molecules (proteins) on the surface of the virus. The precision of the interaction is equivalent to a sophisticated lock and key mechanism, while the attachment of the antibody to the virus can be thought of as setting off a very loud house alarm that signals invasion. Immunologists refer to the bits on the viral proteins that are recognised by the antibodies (the key) as antigens (the lock), and speak about antigen/antibody complexes. The presence of bound antibody neutralises viral infectivity by a variety of different mechanisms, including the attachment of an additional, toxic (for the virus) molecule called complement, signalling to macrophages that the antigen/antibody complex with the attached virus should be eaten and destroyed, or by simply preventing virus entry into the epithelial cells or nerve cells where it likes to grow. In the case of polio, antibodies in the blood stop the virus from getting to the brain and spinal cord. This prevents the paralytic symptoms of poliomyelitis, which result directly from the virus-induced destruction of irreplaceable nerve cells, particularly the large motor neurons.

The antibody molecules are themselves proteins that are made by large production 'factories' called plasma cells. The plasma cells are the progeny of a small subset of white blood cells called B lymphocytes, or B cells, that bear a juvenile form of the anti-poliovirus antibody on their surface and multiply rapidly after vaccination. This process of specific cell division, or clonal expansion, goes on in the lymphoid tissue, the lymph nodes and spleen that provide a 'nurturing' environment for the optimal development of immunity. Any immune response, whether caused by immunisation or infection, is likely to be associated with lymph node distension and effects such as drowsiness resulting from the secretion of various chemicals (called lymphokines) that are normally made by the responding cells. The excess lymphokines spill over into the blood and are then carried to the brain. A small child may be irritable after being vaccinated, and lethargy and 'swollen glands' in the neck are a familiar feature of respiratory infections.

Other lymph nodes in the armpit, the groin and so forth may become enlarged as a consequence of infection in different parts of the body. This can be both spectacular and clinically dangerous in the condition infectious mono-nucleosis (colloquially termed kissing disease) that some-times develops when adolescents are first infected with the herpes virus, Epstein Barr virus (EBV). Infectious mono-nucleosis is an example of an immune response that is, for a time, out of control, though it generally resolves. Almost everyone is infected with EBV and, though most of us live happily with this persistent virus for life, it can cause lymphomas in people who are severely immunosuppressed. Lymphoma, leukosis and leukaemia are all familiar words that describe various cancers, or abnormally enhanced growth patterns, for blood and lymph node cells that do not self-cure. Most of these conditions are not caused by viruses.

After the initial response to poliomyelitis vaccine the plasmablasts, which may be thought of as the more mature and adventurous 'daughters' of the clonally expanded, poliovirus-specific B cells, exit from the lymphoid tissue, enter the blood circulation and eventually lodge in sites like the gastrointestinal tract, lung and bone marrow. Here the plasmablasts change to become the 'terminally differentiated' plasma cells, which don't divide any more but continue to pump out antibodies for years. The presence of poliovirus-specific antibody in blood or gut mucous over the very long term is evidence of immune 'memory'. Checking for this is easy. Blood serum, the fluid phase of blood that remains after clotting, is simply mixed with infectious virus, which is then tested for the capacity to kill tissue culture cells growing in a test tube. If the serum contains neutralising antibody, the cells remain healthy. Otherwise, they become infected, change their shape and then die.

When antibody levels fall it's time to get a booster shot of the vaccine, which re-stimulates poliovirus-specific 'memory' B cells, the 'daughter' cells that stayed in the lymphoid tissue and did not go down the plasmablast/ plasma cell pathway the first time round. These boosted, or secondary, responses are generally bigger than those that occur in an 'immunologically naïve' person (that is, someone who has never encountered either the virus or the vaccine). The reason for this is that the initial response increased the numbers of 'memory' poliovirus-specific B cells way beyond those that would be found in a naïve individual. Most successful vaccines, whether against poliovirus, yellow fever, measles or whooping cough, work essentially as I've described here.

The antibody response that I've discussed so far is referred to as humoral immunity. Most of my work has been on cell-mediated immunity, a different sort of adaptive immunity associated with T cells, which are lympho-cytes manufactured in an organ in the neck called the thymus. Like B cells, T cells bear a single, specific cell sur-face receptor, called the T cell receptor (TCR), but there is an important difference. While the receptor on the B cells is an early form of the antibodies that later circulate as cell-independent proteins in the blood, the TCRs are per-manently cell-associated. In short, the 'effectors' of humoral immunity are the antibodies, but the T lymphocytes with their attached TCRs are themselves the effectors of cell-mediated immunity. Their role is to interact with and then destroy other cells that are modified by, for example, being infected with a virus or as a result of some cancerous change. This function is called immune surveillance.

Broadly, the T cells are classified into two separate categories: the $CD8^+$, or 'killer' T lymphocytes that 'assas-sinate' virus-infected cells, and the $CD4^+$ 'helper' T cells. The term 'CD' is immune-speak, a term used to classify the hundreds of different proteins found on the surface of lymphocytes and other white blood cells. The $CD4^+$ T helpers operate via the secretion of various lymphokines that promote (help) both the clonal expansion and differ-entiation of antigen-specific B cells and $CD8^+$ T cells. The destruction of these helpers by the human immunodefi-ciency virus (HIV) is the primary cause of AIDS: HIV uses the CD4 molecule as a receptor to facilitate its entry into the cell. Once all the $CD4^+$ T help is lost, the collapse of the rest of the immune system soon follows.

The $CD4^+$ T cells have been further sub-divided into Th1 and Th2 sets. Members of the Th1 group can both provide help for some categories of antibody responses and act as 'effectors' in the elimination of viruses and intracellular bacteria. The Th2 cells normally promote the type of antibody response that is meant to control the larger parasites, like round worms. Diseases like multiple sclerosis may be caused by Th1 cells that become abnormally activated against 'self' proteins, while Th2 responses to antigens in, say, cat dander or the excretions of the dust mites that are found in every household, cause potentially fatal conditions like asthma. Much of this research is still evolving and it will be interesting to see whether Nobel Prizes result.

What about the $CD8^+$ T cells, our 'hit men' of immunity? This gets complicated, so I will try to paint a comprehensible word picture. Virus-specific $CD8^+$ T cells can be thought of as resembling those old style sea-mines with spikes all over them. The spikes are the antigen-specific receptors, the TCRs, with each killer T cell bearing about 20,000 identical spikes on its surface. Though they may look primitive, these particular mines are very smart. First, each 'mine' expresses only one of about a million to 10 million different spikes. Again, think of each possible spike as being like a unique key. Furthermore, if our T cell 'mine' first meets the particular 'ship', or virus-infected cell, that it is intended to destroy in the safe harbour of the lymphoid tissue, it will not blow it up immediately but will multiply itself (clonal expansion), then move out through the blood to find and eliminate every vessel that has the same characteristics.

Our $CD8^+$ T cell 'mines' have their own 'motor' system and are very mobile. In addition, they have a flexible casing (cell wall), and can deform and move through various barriers (like blood vessel walls) in pursuit of an intended target. Furthermore, when these T cell 'smart mines' have done their job and all the accessible targets are destroyed, the survivors are programmed to first turn off the various effector chemicals that constitute the 'explosive', then to continue patrolling the blood and organs for years as 'resting', memory T cells. These 'memory mines' will be reactivated, and will respond more quickly if the enemy builds new ships of the same type, that is, if we become reinfected with the same virus.

Thinking about the 'smart sea-mine' T cell, we realise immediately that it would be totally counter-productive for such a sophisticated weapon to recognise, and consequently blow up, our own navy. The reason this does not happen is that the use of any receptors, or spikes, that might recognise the 'self' ships belonging to us is expressly forbidden. This encapsulates the problem of self/non-self discrimination that has been an obsession for immunologists since the earliest history of the subject. The way that this works is that the early-stage T cells that express receptors (spikes) with the capacity to bind self antigens are eliminated during the process of development and differentiation in the thymus: more about that later.

What is it that the $CD8^+$ killer 'smart mines' recognise? Think of each human being as a separate country made up of millions of units, or cells, in some futuristic world. Flags have been relegated to history and national identities are now determined by patterns of big, coloured dots. The code is quite varied and can potentially use hundreds of different colours, with a maximum of six per nation state. These six identical coloured dots are worn conspicuously by every member (or cell) of the population, are on every object identified with that population, and are the primary measure of national pride and uniqueness. Half the dots any one person (nation) expresses come from the mother, half from the father. The dot analogy describes, in fact, the situation for the transplantation molecules, or histocompatibility proteins, that are recognised as foreign (non-self) on a grafted kidney. Our unique dot pattern, or profile of transplantation antigen expression, defines us as individuals, at least in the biological sense. The exception is, of course, identical twins.

Combining the CD8$^+$ T cell 'sea mine/human being-nation' analogies, the spikes (or receptors) on the mines are programmed to recognise the alien dot patterns painted on the bottom of any enemy ships that are suddenly 'transplanted' into our oceans and rivers. Unless they are suppressed in some way, the CD8$^+$ T cell sea mines react violently and sink (reject) the invading ships (transplanted tissues) that carry the foreign, non-self dots. That's fine for warfare, but there's a problem when it comes to biology. The only 'graft' or 'transplant' situation that occurs naturally in mammals like us is pregnancy. The embryo is not the enemy, so why would it be that evolution had favoured the development of a situation where the mother's immune system could potentially 'see' the paternal transplant molecules (dots) expressed in the foetus as foreign, or 'non-self'? Why have this dot system in the first place?

That question was answered when Rolf Zinkernagel and I discovered that immune CD8$^+$ T cells generated during the course of a virus infection recognise infected cells that express our own transplantation molecules, our 'self' dot patterns, that have been modified by the virus so that they now look like 'foreign' or non-self dots. Think of the self dot as now carrying a new 'squiggle', a small piece of novel protein (or peptide) that comes from the invading virus. The uninfected cell next door still expresses the normal, unadorned 'squiggle-free' dots and is ignored. Each of our six self dots, or transplantation molecules, can potentially carry a different squiggle derived from one or other of the virus proteins. The virus can't stop this happening as, during the process of virus replication, the viral genome makes a certain amount of unsuitable, damaged protein that cannot be used to build new virus particles and is thus chopped into bits (peptides) by the shredder system (called the proteasome) that gets rid of excess proteins within the cell cytoplasm. The reason that evolution has favoured the emergence of so many variant dots (a phenomenon known as genetic polymorphism) is that this diversity makes it much less likely that any particular invading virus can sneak past the immune surveillance mechanism by failing to provide a peptide 'squiggle' that binds to (and consequently changes) at least one of the dots.

The so-called transplantation system is, in fact, a self-monitoring alarm system, which ensures that our 'sea mine' T cells will recognise and destroy any 'ship' with altered self dots that has been captured (infected and then squiggled) by the virus enemy and thus threatens the survival of our navy (body). What cell-mediated immunity is all about in the biological sense is getting rid of these 'squiggled' infected cells. Otherwise, more virus particles will be produced, and the extent of both the infection and the resultant clinical impairment will be aggravated. Operationally, the immune system doesn't distinguish between the 'foreign' dots on the surface of cells in a transplanted organ and our own infected cells carrying 'squiggled' dots. Graft rejection is simply a by-product, an epiphenomenon, of this normal self-monitoring process. If this had been discovered first, what we call transplantation molecules would be known as self-surveillance molecules. That doesn't mean that the problem of graft rejection has been ignored through the long march of biology: the mammalian placenta functions to protect the baby from attack by the mother's T cells.

I will discuss the experiments we performed, why they were done and how they were interpreted later in this chapter, but will now finish this short course in immunology. I hope it hasn't been too painful, but immunology is a complicated area of science. A great deal has been left out, but I have tried to provide enough information to allow the following account of the immunology Nobel Prizes to be read with some understanding. What has also been excluded as I go on to discuss the various Nobel Prizes are the names of many great immunologists who, in spite of enormous contributions, did not get to make that particular trip to Stockholm.

The very first Nobel Prize for Medicine to Emil von Behring in 1901 was awarded in the area of adaptive immunity, though it was not called that at the time. He was recognised for the development of serum therapy against diphtheria or, to put it in modern terms, for treating the disease by giving anti-diphtheria antibody. Von Behring's initial experiments used serum from recovered humans, but he then went on to produce larger amounts of antiserum in vaccinated sheep and horses. Anyone who has been injected with tetanus toxoid after a deep cut, or with antivenene following a snake bite, will have received a similar product, though, of course, the toxins against which the particular antiserum was raised are different.

The theme of immune specificity was continued in the 1908 award to Paul Ehrlich. Like von Behring, Ehrlich trained initially with Robert Koch (Medicine, 1905) who was honoured for identifying the mycobacterium that induces tuberculosis. Koch is also widely known in science for formulating Koch's postulates, which define the criteria for accepting that a particular infectious agent is, in fact, the cause of the disease associated with it. Ehrlich, who is commonly acknowledged as the 'father of immunology', was recognised for his insights on the nature of the antigen/antibody (lock and key) interaction. Given all the possible antigens in nature, the spectrum of antibodies is, of course, extraordinarily diverse. He speculated about immune specificity and developed a 'side chain' theory to explain how it is that so many different antibodies can be made. Though Ehrlich identified and stated the immune specificity problem, the mechanism that he proposed was later shown to be wrong.

In thinking about the issue of immune specificity, Ehrlich also recognised the existence of the autoimmune diseases, like rheumatoid arthritis and multiple sclerosis (MS). Autoimmunity, which he described as 'horror autotoxicus', occurs when the immune system gets the wrong message and starts to make T cells and antibodies that react against the individual's own body tissues. Such diseases remain difficult and clinically important medical problems that assume increasing significance with the overall ageing of the human population. In addition, Ehrlich is known for coining the term 'magic bullet' to describe chemotherapy. The year after his Nobel award, the Ehrlich team discovered Savarsan, the first effective treatment for syphilis. Ehrlich died of a stroke in 1915 (and should not be confused with the recently deceased American scientist and author of the same name).

The protective role of antibodies was further elaborated by Jules Bordet (Medicine, 1919), who showed that an additional serum substance, termed 'complement', combines with the antigen/antibody complex to deliver a 'lethal hit' that destroys the cholera organism, *Vibrio cholerae*. We now know that there is a 'complement cascade' involving a number of different proteins, though the credit went to Bordet and the massive body of work over the subsequent eighty years has not attracted a further Nobel award. As a general rule, Nobel Prizes are given only once to a particular sub-field of science. Jules Bordet also established that injecting rabbit red blood cells (RBCs) into guinea pigs caused the formation of antibodies which, acting with complement, eliminated the foreign RBCs. The guinea pig had, in fact, dealt with the rabbit RBCs in exactly the same way that it would handle an invading micro-organism. The realisation that RBCs provoke an immune response when injected into other individuals was in accord with Karl Landsteiner's immensely important discovery of human blood groups that was recognised by the 1930 Medicine Prize.

Arne Tiselius was awarded the 1948 Chemistry Prize for using an electric field, the revolutionary technique of electrophoresis, to separate the various components of serum, including the gamma-globulin fraction that contains the antibodies. Further advances in protein biochemistry and protein sequencing (Fred Sanger, Chemistry, 1958) then allowed Rodney Porter and Gerald Edelman (Medicine, 1972) to explain the molecular basis for the enormous diversity of the antibody 'repertoire' by defining the chemical structure of the gamma globulins and the antibody binding sites. Their work killed off both Ehrlich's side-chain hypothesis and later speculations by Nils Jerne (Medicine, 1985). The opening paragraph of Rod Porter's Nobel lecture is a good example of how a scientific lineage works:

> In 1946 when I was starting work as a research student under the supervision of Dr F. Sanger, the second edition of Karl Landsteiner's book *The Specificity of Serological Reactions* reached England. In it was summarised the considerable body of information available on the range of antibody specificity and much of it was Landsteiner's own work or by others using his basic technique of preparing antibodies against haptens (small, defined antigens) and testing their ability to inhibit the precipitation of the antisera and the conjugated protein. Also described in this book was the work in Uppsala of Tiselius and Pedersen in collaboration with Heidelberger and Kabat (at Columbia University, NY) in which they showed that all rabbit antibodies were in the γ globulin fraction of serum proteins and that they had a molecular weight of 150,000. This combination of an apparently infinite range of antibody combining specificity associated with what appeared to be a nearly homogeneous group of proteins astonished me and indeed still does.

Rod Porter's account mentions the enormously influential immunochemists Elvin Kabat and Michael Heidelberger, who missed out on the Nobel accolade but lived to be 86 and 103 respectively and thus did well in the immunology longevity awards.

Nils Jerne and his colleagues Alastair Cunningham and Al Nordin used Jules Bordet's RBC immunisation model to complete some of the first single-cell studies of the antibody-forming cells, the plasma cells that constitute the final differentiation step in the B cell pathway. The 'B' designation reflects that, in the chicken, the precursors of these antibody-forming cells are made in the specialised Bursa of Fabricius, named for a University of Padua anatomist, a contemporary of William Harvey who worked out the role of the heart and the blood circulation. The Bursa is located near the cloaca, the common opening of the avian digestive and urinary tracts. There is no comparable organ in mammals like us, and the precursors of the B cells are made first in the foetal liver, then in the bone marrow.

The single cell antibody analysis of Cunningham et al added to the accumulating body of evidence that each plasma cell produces antibody of only one specificity (a single key type in the lock and key analogy), a result that was predicted by F. M. Burnet's clonal selection theory of acquired immunity. Burnet shared the 1960 Nobel Prize for Medicine, though it was not awarded for clonal selection, which he evidently regarded as his main contribution to immunology. Mac Burnet had an enormous influence on immunological theory through the 1960s, and his ideas about clonal selection and immune tolerance (see below) continue to define much of the thinking in this area of science. Though I spoke with him briefly on only two occasions not long before he died in 1985, his books on both virology and immunology greatly influenced the career path that I took as a young veterinary science graduate. He told his own life story in the 1968 book *Changing Times: An Atypical Autobiography* (Heinemann) and there is an excellent 1991 biography by Christopher Sexton, *The Seeds of Time: The Life of Sir Macfarlane Burnet* (Oxford University Press).

Nils Jerne's co-recipients of the 1985 Medicine Prize were Georges Kohler and Cesar Milstein. They took Burnet's clones of antibody-forming cells, and 'immortalised' them as 'hybridoma' cell lines that can be used indefinitely to make enormous amounts of antibody of a single specificity. The availability of these 'monoclonal antibodies' (mAbs) revolutionised many areas of biology because they provide precise, reproducible reagents that bind very specifically to structurally defined antigens. Most of us now buy these mAbs from commercial suppliers. Like the manifestation of Mab the Queen of the Fairies in *Romeo and Juliet*, the mAbs are both small and magic! Injected mAbs are also proving to be increasingly useful for cancer therapy.

Perhaps the last chapter in the Nobel Prize antibody story came with the 1987 Medicine Prize to Susumu Tonegawa, 'for his discovery of the genetic principle for generation of antibody diversity'. Susumu explained how the genes coding for the antibody molecules organise to produce such an extraordinarily varied spectrum of functional proteins with different specificities. If there is another Nobel award for the antibody area, my guess is that it might recognise some enormously important discovery or application that relates very directly to clinical medicine. Possible candidates would be involved in the areas of autoimmunity or cancer immunotherapy, or in the creation of a new basic principle or technology allowing the development of effective vaccines against diseases like TB, HIV/AIDS or malaria.

Our 1996 award was 'for their discoveries concerning the specificity of the cellular immune defence'. This is the latest but not, I hope, the last Nobel Prize for illuminating cell-mediated immunity. Again, as with the antibody studies, there is a lineage that can be traced back via the Nobel awards.

Many immunologists believe that the one obvious accolade that has been missed from the immunology Nobel list is a prize for discovering the basic function of the thymus. The thymus remained a major mystery from historical time through the first half of the twentieth century. The breakthrough came at the beginning of the 1960s when a young Swiss-born Australian, Jacques Miller, who was then working at the Chester Beatty Cancer Institute in London, removed the thymuses (thymectomy) of baby mice as part of a study he was doing on a virus-induced leukaemia. The unexpected, serendipitous result was that, as they matured, the thymus-less mice failed to develop T cells and thus an effective immune system. At about the same time, Robert Good and a young associate at the University of Minnesota, Max Cooper, found similar effects with thymectomised baby chickens.

The Nobel Prizes that have been given for discoveries and technological advances related to T cell-mediated immunity are all, in one way or other, related to the broad area of transplantation and tissue graft rejection. Continuing the theme of self/non-self discrimination that originates with Ehrlich's 'horror autotoxicus', Macfarlane Burnet and Peter Medawar were awarded the 1960 Medicine Prize 'for discovery of acquired immunological tolerance'. The Melbourne-based Mac Burnet proposed the theoretical basis of what is referred to by immunologists as 'tolerance', while Peter Medawar in England did some of the key experiments establishing that the rejection of foreign (non-self) skin grafts is indeed mediated by a specific immune mechanism. Tolerance means that the immune system has to develop in such a way that it will react against, and eliminate, a non-self (foreign) invading virus or parasite, while at the same time ignoring (tolerating) self. Many immunologists remain focused on this problem of self/non-self discrimination. When it comes to T cells, that means telling the difference between our own, normal 'dots' (transplantation molecules) in the T cell 'smart mine' analogy outlined above and both the foreign 'dots' on the cells of other individuals and the 'squiggled' dots of self cells infected with a virus.

Tolerance works in such a way that the precursors of the immune cells 'learn', as they develop in the thymus, to avoid reacting against self, the cells expressing the particular spectrum of 'dots' that is normal, but different for each and every one of us. Think of the thymus as a severe preschool environment for T cell development: any of these cells that bears a receptor reactive to our own (self) proteins is simply eliminated before it can enter the blood. The thymus is a tough pre-school, a bit like the Spartans who dropped defective children off the top of a cliff. This is reminiscent of the obscene philosophy of social Darwinism and we should be very leery of translating biological themes into social practices.

Though it may seem severe that the thymus functions as both a T cell judge and executioner, selective cell death is a normal part of all development. Many cells die as an organ assumes the familiar shape of a brain, liver or a kidney. That story was worked out by Sydney Brenner, Bob Horvitz and John Sulston from the study of a tiny round worm called *Caenorhabditis elegans*, leading to their 2002 Nobel Medicine Prize. This strategy is referred to as 'altruistic suicide'. Without death there can be no enduring life. All cells are programmed to die, and will normally do so if they are excess to requirements or damaged in some way. One way that cancer develops (see chapter 8) is when this suicide process fails in a cell bearing a dangerous mutation. The so-called 'killer' T cells that I study are actually 'accessories before the fact' of suicide. They function by transmitting molecular signals that tell any virus-infected cell they recognise to commit suicide. In turn, some viruses have evolved mechanisms that are intended to defeat this normal elimination mechanism.

The reason that we can't simply inherit the rules for self/non-self discrimination is that we receive our genetic material at random from two quite different parents, who express varied spectra of the six different transplantation molecules (dots). The T cell learning experience thus has to be repeated during the development of each individual human being, frog, mouse or chicken. Like any form of education, the selection process in the thymus is imperfect because T cell-mediated autoimmunity can develop in later life, but that requires special circumstances that are too remote from the main theme to be discussed here.

Following Burnet and Medawar in 1960, the next Nobel Medicine award for work in the general area of cellular immunity was in 1980 to Baruj Benacerraf, Jean Dausset and George Snell (from Boston, Paris and Maine, respectively) 'for their discoveries concerning genetically determined structures on the cell surface that regulate immunological reactions'. The citation more accurately describes the work of Benacerraf than that of Snell or Dausset. What the other two did was to establish the basic rules governing the inheritance and character of the trans-plantation antigens. I describe below how Snell worked out the story experimentally in the mouse, while Dausset dis-covered the comparable human lymphocyte antigen, or HLA, transplantation system when he found that indi-viduals who had been given many blood transfusions also developed antibodies against white blood cells. He realised that he was looking at something very different from Landsteiner's blood groups.

Though it remains a recurrent cause of confusion in the media, the blood groups define only the rules for red cell (erythrocyte) transfusion and are quite distinct from the transplantation molecules found identically on the skin, the kidney or, for that matter, the white blood cells. Human beings are quite sophisticated in their understand-ing of many things, but it is extraordinary how ill-informed even the well educated can be when it comes to the func-tioning and nature of their own bodies. A positive feature of the Internet is the rise in general biomedical knowledge as people find that they can research their own questions and get satisfying answers. Even so, the grievous errors made all too regularly in medical and science reporting are just one of the reasons why all responsible media outlets should employ at least one well-qualified science journalist, a plea

I made earlier in this book. Maybe some could do with just one fewer sports writer!

At the beginning of his career in the 1930s, George Snell decided that he would investigate immunity to cancer. He found that tumours transplanted from one mouse to another were soon eliminated, but quickly realised that these rapid responses were directed at what came to be called the histocompatibility (H2), or transplantation, antigens (the 'dots') characteristic of the different individual mice rather than at anything specific for the tumour. His next step was thus to develop inbred mouse strains so that the donors providing the transplanted tumours would be genetically identical to the recipients that might then, it was hoped, develop a tumour-specific immune response. However, as happens so often in research, he soon came to the conclusion that the more important job for him was to work out the mouse transplantation system and to leave the cancer immunity problem to the next generation. The average 'generation' time in science is about ten years. Snell didn't have even the remotest idea that what he started would also break open the whole story of how immunity works to defeat infection: our experiments that came much later depended totally on what he and his colleagues had done.

The systematic back-crossing (brother–sister matings) approach that Snell used to develop mouse strains capable of accepting reciprocal tail skin grafts from any one individual to another was greatly facilitated by the fact that the inbreeding job had already been half-done much earlier in the twentieth century. A Massachusetts school teacher and 'hobby' farmer, Miss Abbie Lathrop, had to retire early because of ill-health, and turned to breeding ornamental mouse strains as a means of paying her bills. Like Cruft's dog show today, exhibiting 'fancy' mice was a popular hobby in the late nineteenth–early twentieth century. If Snell is the father, Miss Abbie can legitimately be regarded as the mother of modern mouse genetics, though I doubt that they ever met. There are plenty of photographs of George, but the only image of Miss Abbie that I've seen is an engraving that shows her looking pensively at a blob of fur on top of a wire mouse cage. The C57Bl/6 strain that we use in most of our mouse experiments today in both Melbourne and Memphis is in a direct lineage from one of Miss Abbie's 'fancy' mice. I wonder what she might have thought if she could have known that she started what became an enormous enterprise in medical discovery and defeating disease.

The way to 'fix' a desirable trait in any species from cats to cattle is to do brother–sister matings. That's why a beef Hereford looks like a Hereford and not like a milking Jersey. The downside, however, can be, as anyone who owns a dachshund (spinal disc problems) or an Irish setter (retinal atrophy) knows, that selecting aggressively for one thing can lead to other undesirable consequences. In medicine, of course, this can be an advantage, as an inbred mouse that develops an abnormality characteristic of a particular human disease provides an extraordinarily helpful model for studying the problem. 'Mega Mouse', the Jackson Laboratories in Bar Harbor, Maine, where Snell spent his career, has provided the research world with an enormous variety of such disease-prone mouse strains. The attendants who looked after the mice, cleaned the cages and so forth, were trained to keep any sick-looking individuals that turned up in a litter and bring them to the attention of the Jackson Laboratory scientists. Bar Harbor, on the beautiful Mt Desert Island, was (and is) a great place to work for those who don't mind a degree of isolation, especially in winter. George Snell is quoted as saying: 'I was on a hunt for 30 years. I wore a laboratory gown, not a Maine guide's red wool jacket'.

Recently, the development of transgenic/knockout mouse technology has meant that this process of throwing up natural disease variants can be short-circuited by disabling the particular mouse gene in very early embryos, then popping in the comparable, abnormal human gene. The availability of such genetically modified mouse lines has made it much easier to determine how, for instance, a mutated gene promotes the development of cancer. Most of us expect that there will be a Nobel Medicine Prize for the science that led to the transgenic/knockout mice; they have been enormously important in all areas of biomedical research, including our own studies of viral immunity. We can, for instance, now do virus infection experiments in mouse strains that selectively fail to make antibody, $CD8^+$ T cells or $CD4^+$ T cells, providing powerful analytical tools for sorting out what these various components of the immune system actually do. The Jackson Laboratories serve as a worldwide repository for these extraordinarily valuable mouse strains.

Going back to earlier, simpler times and the old genetic technology of inbreeding and back-crossing, George Snell and his friends and competitors used conventional mouse strains as a starting point to develop genetic recombinants, separating the three different regions (called loci) in the mouse H2 gene complex that are associated with rapid skin graft rejection. Comparable HLA genetic loci have also been identified in humans. Knowing about the HLA system has allowed us to define how human T cells work, and informs research in every field from HIV/AIDS, to cancer, to autoimmunity to organ grafting. Without the mouse studies, sorting out both the details and the associated functions of the human HLA transplantation system would have been extremely difficult. The laboratory mouse is one of the heroes of medical research.

Working initially with outbred guinea pigs, Baruj Benacerraf found CD4$^+$ T cell responders and non-responders to a synthetic antigen made by the chemist Paul Maurer. This was the first instance of an immune response that looked as if it was under the control of a single dominant gene. As often occurs, serendipity stepped in and Baruj benefited from the fact that the great population geneticist and evolutionary theorist Sewall Wright just happened to have developed two inbred guinea pig lines that allowed him to define his immune response (Ir) gene effect. Using mice rather than guinea pigs, Stanford University's Hugh McDevitt soon showed that comparable Ir genes were located in the H2 complex, though not at the loci associated with rapid graft rejection. The analysis then proceeded much more rapidly with the mice, as mouse genetics provides a more precise and defined analytical tool. Being a smart scientist, Benacerraf quickly switched to the superior mouse model system.

Most immunologists were grateful that the award of this 1980 Nobel Prize was somebody else's responsibility, because the rule of three meant that other major players were excluded. The English scientist Peter Gorer, who used the antibody response to define the mouse H2 system, was not in contention, because he died too soon. Apart from McDevitt, the Dutch HLA geneticist Jan Van Rood was also considered by many to have a strong case. In these situations, the decision is likely to be made essentially on the basis of priorities defined by publication, which is why it is unwise to hold back a major discovery from publication.

The clinical transplanters Joseph E. Murray and E. Donnal Thomas were then recognised by the 1990 Medicine Prize 'for their discoveries concerning organ and cell transplantation in the treatment of human disease'. Joe Murray was the first to do successful human kidney transplants, and Don Thomas pioneered bone marrow transplantation. The motivation for the latter was to replace the immune and blood-producing systems of cancer patients who had been treated with massive doses of chemotherapy and radiation. Organ transplantation is now, with modern immunosuppressive drugs, a routine procedure, and bone marrow reconstitution is a standard protocol in major cancer hospitals.

Our prize, which came exactly 200 years after Jenner gave his cowpox virus to James Phipps, was for work that Rolf Zinkernagel and I did trying to develop a better understanding of the cellular immune response to lymphocytic choriomeningitis virus (LCMV). This is a fascinating virus that causes a persistent, inapparent infection with no clinical consequences in neonatal mice that have not yet developed an effective immune system, but can be lethal when given to immunocompetent adults. Before Rolf arrived in Canberra, I had been following an idea, then prevalent in the LCMV field, that the elimination of virus-infected cells in the central nervous system (CNS) by immune T cells also kills the mice. This T cell-mediated destruction is called immunopathology. It could be said to be the equivalent of the we-had-to-destroy-it-in-order-to-save-it concept that became familiar during the Vietnam War.

The levels of protein in blood are normally much higher than those in the cerebrospinal fluid (CSF) that bathes the brain, but protein floods into the CSF when the LCMV-immune T cells kill the virus-infected cells in the blood brain barrier that maintain this differential. The resultant increase in CSF protein concentrations (osmotic imbalance) draws more water into the CNS, which in turn causes acute brain swelling. This is the same effect that can kill humans when they suffer severe head injuries in, say, a car crash, when an acutely swollen brain is compressed by the bony confines of the skull. The focus of these LCM experiments reflected my experience during the preceding five years in Edinburgh where, as a neuropathologist, I had been working on diseases of the brain and spinal cord.

Rolf came to Australia to work on bacterial immunity with a close colleague, Bob Blanden, but, as there wasn't room for him in Bob's laboratory, he ended up sharing with me. Rolf also tends to be rather noisy and loves opera, which is a long way from Bob's musical taste. At that stage, I remember him rendering (basso, but not too badly) Cherubino's song from Mozart's *Marriage of Figaro*, which is usually sung by a girl. As I'm also an opera and classical music fan, we got along fine and still do, though we don't always agree scientifically. A Basle Swiss, Rolf had taken an immunology course with the great (and very modest) Zurich scientist Jean Lindenmann, one of the discoverers of interferon; he then moved to Lausanne, where he worked for a relatively short time with another famous Swiss, Henri Isliker, on bacterial immunity in rabbits. Much later, Rolf was to succeed Lindenmann as Zurich's leading immunology figure.

Way back then, we talked, and decided to collaborate by combining the LCMV model with the relatively new cytotoxic T lymphocyte (CTL) assay that he had used previously in Lausanne. Though they weren't called that until later, this CTL assay measures CD8+ T cell effector function. Our first experiment analysing the 'meningitis' cells recovered from the cerebrospinal fluid of LCMV-infected mice was a spectacular success. As predicted, very potent, 'effector' CD8+ T cells had exited the blood and localised to the brain where, just as they killed virus-infected tissue culture cells in the little glass tubes of our CTL assay, they were clearly capable of damaging virus-infected CNS cells. This was our first joint paper, published during late 1973 in the Rockefeller University's prestigious *Journal of Experimental Medicine*.

A leading scientist in the LCMV field, Michael Oldstone from the Scripps Institute in La Jolla, had published the results of some collaborative experiments with Hugh McDevitt and Graham Mitchell, one of Jacques Miller's graduate students who was then working with McDevitt at Stanford. Their paper suggested the possibility of an H2-related 'Ir gene effect' in the LCMV immunopathology model. We decided to follow up this idea with our CTL assay, and got hold of a couple of H2 different mouse strains that were being used for graft rejection studies. These were infected with LCMV and analysed for the possibility that there might be H2-related differences in the magnitude of the CTL response.

Much to our surprise, we found that the LCMV-immune cells killed only the virus-infected cells that shared the H2 antigens of the infected mouse. The excitement was enormous. We had been reading some of Baruj Benacerraf's papers and thought initially that we had in some way replicated the type of effect that he was then studying. Even if that had been the case, it would have been a big finding because our work involved the 'real world' of viruses rather than synthetic antigens. As it turned out, we had discovered something quite different and totally unpredicted. The first paper, which you will find in Appendix 1, was published promptly in *Nature* in early 1974, with the help of the leading English immunologist, John Humphrey, who acted as a conduit to the editors. A second quickly followed (Appendix 2). In all, a total of four such short letters from our collaborative efforts appeared in *Nature*.

We rapidly followed up this serendipitous finding with all the mouse strains that we could lay our hands on. The one genetic recombinant (called A/J) that happened to be available in Canberra allowed us to map the T cell recognition effect to the loci associated with acute graft rejection. These A/J mice were being used by the then professor of zoology at the ANU in a big experiment designed to look at the 'natural' evolution of mouse populations. One of the graduate students 'alienated' a few 'spare' mice from his breeding colony (not the experimental group) for our study. I don't know if he was ever aware of this. Minor criminality is deeply embedded in the Australian experience. Australia was, after all, initially founded by a bunch of criminals, some of whom were convicts while others were among those who wore military uniforms. Anyway, it was a key experiment and those few 'borrowed' A/J mice blew the story wide open. This was another *Journal of Experimental*

Medicine paper. Many experiments followed over the next 12–18 months, particularly with the spectrum of H2 genetic recombinants generated by Snell and his intellectual descendants in the United States.

This A/J analysis mapped our effect to regions of the mouse H2 gene complex that were different from those studied by Benacerraf, McDevitt and their colleagues. The result was enormously exciting, as it was immediately clear that we had found something radically new. The results were solid, and the field of transplantation was turned on its head. The key biological role of the H2 antigens was not to be 'seen' as 'foreign' on a grafted tissue, but to be the target for some self-monitoring, or immune surveillance, function concerned with eliminating abnormal cells within, say, a virus-infected person. This solved a major mystery in biology. Up until that time, as I said earlier in this chapter, there seemed no good biological reason for the phenomenon of graft rejection. Luminaries like Mac Burnet, Nils Jerne and George Snell had all speculated about it, but none had been able to guess the right answer.

As we were looking at virus-infected cells, and as viruses subvert normal cell functions, the thought that immediately came to mind was that the virus was in some way modifying the self H2 antigen. We thus proposed that our CD8[+] T cell assassins could be using a single T cell receptor (TCR) to interact with 'altered self', defined either as a complex of the virus and H2 or some virus-induced change of the H2 transplantation protein (Appendix 2). The Ir gene effects found by Baruj Benacerraf and Hugh McDevitt for CD4[+] T cells might simply be explained by the fact that some 'foreign' elements were simply unable to make an appropriate association with the particular H2 or

HLA molecule. Most importantly, we had explained why it is that effector CD8$^+$ T cells are targeted to infected cells by recognising the modified transplantation molecules on their surface rather than to free virus particles which, of course, lack these structures. This is the basis of the 'cellular immune defence' identified in the Nobel citation. Rolf and I worked together for less than three years, but we published a few joint papers and reviews after that. The experience was truly intense, as we were totally obsessed and driving our experimental program night and day. It was very hard on Penny and Katherin, who were juggling jobs of their own, caring for small children and managing two insane spouses. Rolf and Katherin's son Martin was born during that time in Canberra, and Penny helped greatly in looking after Annelies and Tini, while at the same time caring for our boys Jim and Mike, all of whom were under school age. Though we had always taken family photographs, there are almost none from that period. On reflection, Rolf and I were obviously pretty impossible to be around. It is a tribute (as J. G. Ballard would have it) to the 'kindness of women' that both couples are still together.

Also, it can't be the greatest fun in the world for any established scientist to be sitting next door to two young, noisy, and not totally modest, guys who have found something really big. We used lots of resources and dominated the limited facilities for gamma counting that formed the read-out for our experiments. Intellectually and experimentally, everything moved very fast. Rolf did the lab studies, while my responsibility was the writing and the mouse experiments. As I recall, we were at times both secretive and communicative, and were certainly obsessed. Fortunately, we were in an institution with a culture that respected our independence, where we were able to drive our own research without interference. Looking back, it was a very special time in a special place. Such experiences don't go on for long and, though there have been many other high points and lots of intellectual excitement in a long research career, nothing for me has ever equalled that time.

The 'single-TCR/altered self hypothesis' that we proposed in the mid-1970s proved to be both heretical and unpopular. Most immunologists believed for many years that there were two TCRs, one for the viral component and one for the H2 antigen. Some did experiments that seemed to prove the two-receptor idea. When the truth eventually was established, a few were destroyed by the fact that they had published wrong information, while others just shrugged their shoulders and moved on to continue highly successful careers. Immunology is such a complex field that there are no particular penalties for misinterpreting the nature of a biological mechanism, especially when the ideas are based on immensely difficult experimental systems.

The proof that the 'single TCR/altered self' idea was right came in the mid-1980s, when Mark Davis, Tak Mak and others established that T lymphocytes indeed express a single, two-chain TCR that is organised much like an antibody molecule. Will this be a future Nobel Prize? Alain Townsend did his experiments showing that the key 'non-self' components (the squiggles) are peptides derived from viral proteins. Further confirmation came when the Harvard structural biologists Pam Bjorkman, Don Wiley and Jack Strominger reported the presence of a 'smear' equivalent to Townsend's 'non-self' peptide in the exposed 'groove' of an HLA molecule (the dot). The final definition of the 'single

Killer T cell

Virus-infected cell

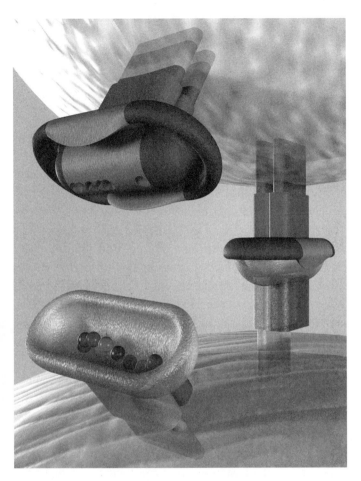

Both figures reproduced courtesy the Australian Academy of Science;
© Jan Schmoeger, Designpoint Pty Ltd

TCR/altered self' model emerged in 1996, the year we won the Nobel Prize, when Ian Wilson at the Scripps and Don Wiley at Harvard published their three-dimensional pictures derived from the structural analyses of TCR-peptide/H2 co-crystals. The nature of T cell recognition is illustrated on the poster designed by the Australian Academy of Science to explain the basis of our Nobel award to a broader, non-specialist constituency (pp. 128–29).

The research lineage that led to our 1996 Prize is fairly typical: the individuals who were recognised grew up in an intellectual and technological framework established by those who went before. I was intrigued early on by reading Mac Burnet's books on immunity and infectious disease, and Jacques Miller, Rolf Zinkernagel and I worked in the Australian 'school' of immunology that Burnet founded. Most of the prizes over the years were awarded for dis-coveries made during the course of experiments designed in the context of earlier ideas, and there are, of course, elements of serendipity in many of the individual stories. If it hadn't been for the mouse geneticists like George Snell, Rolf and I would have had little to work with and the understanding of T cell-mediated immunity could have been delayed for a decade or more. Also, though the indi-viduals who make a discovery are the ones honoured by a Nobel Prize, the development of a particular finding to the point that it becomes both accepted and incorporated in the normal science discourse often depends on an enor-mous amount of work done later by others.

Since its inception in 1901, the Nobel Prize has gone to seventeen individuals who could reasonably be described as professional immunologists. The big international immu-nology meeting held every three years commonly brings together more than 5,000 people. This is a very dynamic field, and a question worth asking is what immunology has done for humanity over the past hundred years or so and where is it now going.

The biggest triumph is undoubtedly in the area of vaccines. Family sizes in the advanced countries at the end of the nineteenth century were much larger than they are today. One of the reasons for this is that the survival of a child was much less assured, as so many died from what are now preventable diseases. Others were compromised with paralytic poliomyelitis, or the deafness and so forth that could follow measles. Some teenagers were lost to the appalling subacute sclerosing panencephalitis (SSPE), a disease that results from the re-emergence of a defective form of the measles virus that hides in the brain then reactivates years later to cause paralysis and death. SSPE is now essentially unknown in countries that have good vaccine coverage. Providing parents in the West do not have some religious or other objection to vaccination, they will not lose an infant to diphtheria, whooping cough, poliomyelitis or measles. The failure to vaccinate raises, of course, major questions concerning beliefs and responsibilities in society, especially as the protection of, say, young babies who cannot be given measles vaccine depends on the 'herd' immunity that ensures this virus is not circulating in the population. The fall-off in vaccine coverage to below 50 per cent in the newly independent states of Eastern Europe after the collapse of the Soviet Union led to a diphtheria epidemic.

Having related some of the triumphs, it is also clear when discussing the vaccine field that we have only solved the easy problems. New vaccine technologies, like the reverse genetics approach that will allow the more rapid production of influenza vaccines, are emerging, but they are simply improved versions of established products. The real challenge for immunology is that, though diseases like malaria and tuberculosis kill hundreds of thousands each year in the developing world, we still have no effective vaccines. The more recent horror of HIV/AIDS is destroying millions of lives. All these infections share the characteristic that the inducing agent 'hides' and persists in the face of a partially effective immune response, which may later collapse. We can recognise the problem, but it is likely that some major conceptual and technical breakthroughs will have to occur before we find the solution.

Most people would find it difficult to think in terms of a world without matched donor blood transfusion, a practice begun as a result of the work of Karl Landsteiner in the first quarter of the twentieth century. Though there are historical accounts of attempted organ and tissue transplantation that go back to ancient Egypt, the successful kidney, heart, lung and liver transplantation that many take for granted today developed with the progressive illumination of the nature of immunity, especially cell-mediated immunity, over the past fifty years. The Holy Grail in the transplant world is to learn how to make people tolerant of grafted tissues and organs without the need for continuous treatment with immunosuppressive drugs. The US National Institutes of Health are, for instance, funding a substantial 'Immune Tolerance Network' in an effort to find solutions. It should be possible, but we're certainly not there yet.

Another question that is being actively researched in the transplant field concerns the possible use of organs from other species. There are simply not enough human organs to go around. Using genetic engineering approaches, pigs are being modified so that they no longer express molecules on the surface of their cells that induce a very rapid rejection reaction. This is not an immune-mediated response, but simply reflects a basic molecular incompatibility between pigs and humans. One concern in this area of research has been the possibility that some hidden pig virus may transmit with an organ transplant and, like HIV (which came from chimpanzees), become established in the human population. My sense is that this is highly unlikely, though the situation would need to be very closely monitored. So, if pig-to-human transplants ever become feasible, it will be intriguing to see how the tension between belief and survival plays out for those who aren't too fond of pigs.

Influenza

The worst single infectious disease catastrophe of the twentieth century was the extraordinarily lethal influenza pandemic of 1918–19. Somewhere between 20 and 100 million people died, the wide variation in estimates reflecting the fact that accurate counts are not available for what is now called the developing world. At the time of writing, there is considerable trepidation that we could be in for something even worse. Influenza viruses spread very rapidly and, with jet planes rather than ships and more than three times as many people on the planet, this is a disaster waiting to happen. John Barry's recent book *The Great Influenza* gives a very readable and engrossing account, not only of the events of 1918–19, but also of the considerable political and social consequences.

The problem is this. The influenza A viruses that cause most cases of genuine 'flu infect a wide range of species, from birds to horses, to seals, to pigs, to humans and so forth. Though they can cross from one species to another, they do tend to be most infectious for their 'maintaining host'. For some years now, the Asian countries have been experiencing a massive, and highly lethal, outbreak in birds of an avian influenza virus called H5N1. The 'H' stands for haemagglutinin and 'N' for neuraminidase, the two surface proteins of the 'flu viruses that are recognised by the protective neutralising antibodies elicited by the standard vaccination procedures. These H5N1 viruses have transmitted naturally to people, causing death rates of greater than 50 per cent. However, there has been little spread from person to person.

Why are we so concerned? The genetic material of the influenza viruses is inherited as eight distinct segments. If, for example, the lung cells of a human or a pig were to be simultaneously infected with an 'avian' H5N1 virus and a standard 'human' H3N2 virus (the familiar Hong Kong 'flu) we could end up with a new 're-packaged' virus that had some elements from the avian, some from the human strain. The possible flash point with H5N1 is in countries like Vietnam and Cambodia where there is little vaccine coverage against the human virus and large numbers of pigs and people living in relatively close proximity to H5N1-infected ducks, the highest-level virus shedders.

An H5N1 virus with 'human' growth and transmission characteristics could be immensely dangerous, as none of us have pre-existing antibodies to H5 or N1. This has probably happened many times before: the H3N2 virus that emerged in 1968 originated from a duck virus. Variant H3N2 viruses generated as a consequence of the selective pressure applied by neutralising antibodies (a process called 'antigenic drift') have been causing pandemics every two years or so since 1968, commonly killing between 20 and 40,000 Americans in a typical 'flu season. Anyone who doubts Darwinian evolution can see it constantly at work in both 'flu viruses and HIV/AIDS.

What can we do? The main problem is that we don't know what 'shifted' variant of the H5 viruses will emerge in humans. This may be a case, though, where 'close enough' may give sufficient protection to save lives, though it may not stop people from becoming sick. A new 'reverse genetics' strategy is being used to pop out the H and N genes from the standard laboratory strains, substitute likely H5 and N1 variants, then use these to make the standard formalin inactivated 'flu vaccine that is familiar to most of us. The difference is, however, that these are genetically modified organisms (GMOs), so they are currently going through new licensing processes as this will be the first such viral vaccine for human use.

Even when these new vaccines are approved, however, the question will be whether we can get adequate supplies of the product out there fast enough. The other protection we have is the so-called designer drugs, Relenza and Tamiflu, which target the influenza N protein. Many nations are stockpiling Tamiflu. Australia led the charge on this. However, the combination of cost, availability and uncertainty about whether the event will actually occur makes it unlikely that most countries will accumulate enough drugs to protect more than those in the essential services, particularly the medical professionals.

Science has thus given us the means to combat a disastrous influenza outbreak using technologies that were not available in 1918–19: that virus was not even isolated until 1933. Nonetheless, the logistics of first making and then distributing the 'right' vaccine could mean, according to a recent WHO estimate, that now humanity might well sustain some seventy million-plus deaths worldwide before appropriate measures are in place. We live with uncertainty on this issue.

The monoclonal antibodies, the magic mAbs, have proven to be extraordinarily valuable. They are now used in all sorts of biological standardisations and in a broad spectrum of rapid, simple diagnostic procedures. Many of these tests employ variations of the types of immunoassays first developed by Rosalyn Yalow (Medicine, 1977) to measure very low concentrations of hormones in blood. Much of my own research uses mAbs to identify various proteins on the surface of immune T lymphocytes that allow their characterisation and separation by flow cytometry. Efforts are being made to bring the mAbs and electronics even closer together, with the possible development of nano-devices (chips and so forth) that come partly from the world of biology, partly from the physical sciences. These will be mini bio-machines.

Now, following the path pioneered by von Behring and his antibody treatment of diphtheria a century ago, 'humanised', genetically engineered mAbs are being increasingly used for cancer treatment. Products like Herceptyn, Rituxan and Avastin are constantly emerging on the market. This area is likely to develop enormously over the next few years. Also, once the nature of the interaction between the mAb and the particular tumour is understood at the molecular level, the hope is that then a small molecule 'mimic' can be designed and produced using chemistry rather than biology. This is a key focus in the area of rational drug design, which is discussed in chapter 8.

Immunology has allowed us to recognise autoimmune diseases like multiple sclerosis, systemic lupus erythematosus (SLE, or lupus), myasthenia gravis (meaning grave muscle weakness) and so forth for what they are. We know, for instance, that myasthenia gravis is caused by autoantibodies against the acetylcholine receptor that is involved in neuromuscular transmission, while the characteristic skin roughening and reddening of lupus is due to the deposition of 'self' antigen/antibody complexes. Having that information does not mean, though, that these disease problems are in any sense solved. Using mAbs to neutralise tumour necrosis factor, a lymphokine secreted by auto-immune CD4$^+$ T cells, has proven to be valuable for treating some individuals suffering from rheumatoid arthritis, though it is not so clear why this approach fails to work in others. Despite enormous effort, we still do not understand which particular brain component is being targeted by the immune T cells that cause multiple sclerosis. There is a great deal of work to be done in the area of autoimmunity.

Cancer immunotherapy with antigen-specific CD4$^+$ and CD8$^+$ T cells is the current focus of a number of active research programs. At St Jude Children's Research Hospital, some patients were massively immunosuppressed for cancer treatment, then given partially matched bone marrow from a related donor to restore their blood-forming cells and, ultimately, their immune systems. The latter process takes months and, in the meantime, a proportion of these extremely vulnerable children developed potentially lethal lymphomas induced by EBV present in the donor bone marrow population. The problem was solved when Cliona Rooney, Helen Heslop and Malcolm Brenner (from Ireland, New Zealand and England respectively) grew up purified EBV-immune T cells from the donors, then gave them to the next group of children to be treated at, or soon after, the time of bone marrow transfer. Broadly comparable approaches have proven to be much less effective for dealing with solid tumours in adults, but there has been a measure of success with diseases like melanoma. The challenge is to work out why some treatments succeed and others fail.

An expanding area is certain to be the immunology of ageing. As we get older, our immune systems decline and we become more susceptible to many infections: older people died at a much higher rate in the recent SARS epidemic. Influenza tends to be most severe in very young children who have never encountered the virus before and in the elderly. Even from age 55, it's wise to take the annual influenza vaccine shot. Influenza used to be known as the 'old man's friend' because it often quietly removed the frail elderly who had come to a very low quality of life. Take a baby with respiratory syncytial virus infection to see Granny, and the consequences for her can be disastrous, even though she will have had the disease in infancy and then been re-exposed when her own children had the disease; over the intervening years, her immune memory has slowly run-down. Given that many more senior citizens are leading happy, effective and productive lives, we need to develop a much better understanding of what promotes and maintains effective immune memory and protection. This is also a focus for vaccine development strategies in general, no matter what the target age-group might be.

Apart from the various medical applications that I've mentioned here, there are also some very basic conceptual questions that continue to intrigue. One of the more perplexing is, how does the immune system maintain a relatively constant size? Unlike the brain, which is limited in its expansion by the tight box of the skull, the immune cells are both dispersed throughout the body and highly mobile. Still, the whole system manages to maintain some sort of balance. The way that immunity is regulated is far from clear, though it is the focus of work in many research laboratories, especially those interested in asthma and the autoimmune diseases.

Given the new technologies and insights that are constantly emerging from fields as diverse as imaging and informatics, there can be no doubt there will be many more discoveries and practical advances in the world of immunology. We can also live in the absolutely certainty that, if this account of the past 100 years in immunology is updated at the end of the twenty-first century, whoever is writing it will be able to say, 'They didn't even understand what many of the questions were then, let alone know the answers'. Biology is extraordinarily complex, and immunology is a particularly complex area of biology. To paraphrase Winston Churchill, we have just reached the end of the beginning.

5

Personal Discoveries and New Commitments

nlike rock stars, Nobel laureates can step out to buy milk and potatoes without being mobbed in the street. Still, they find themselves and their views in considerable demand, and have to decide what really matters. There is never enough time. Several different constituencies claimed me as a 'favourite son' immediately following the announcement of the award, and other worthy groups, particularly the science teachers and communicators, asked for my support. My contact with these more diverse communities has modified my overall priorities and has led to my rethinking a number of issues, a process that continues.

The practice of science is still high on the list of what drives me, and I am as excited by intriguing new data as are my young colleagues who do the experiments and generate the results. I remain 'hooked' on discovery. At some stage, however, I will quit laboratory science and, depending how well my brain is functioning then, devote my remaining time to either staring at the waves or trying to deal intellectually with the 'bigger picture'. This would be an easier decision if I felt that I did not still have some useful role to play in research teams working on AIDS and influenza, both of which are pressing human problems. Still, I have also long held the view that there is a point where senior scientists should metaphorically hang up their pipettes and, like the emperor Cincinnatus, who cleaned up the Roman state and then quit, rusticate themselves. Nobody is indispensable, and I regularly ask my colleagues to let me know when it is becoming obvious that I have passed my use-by date. Senior scientists (like politicians) sometimes go on too long and lose their sense of perspective: age can play bad tricks on what have been incisive minds.

Although I no longer head the department of immunology at St Jude Children's Research Hospital in Memphis, I still maintain a small, but active, laboratory there and enjoy the many contrasts with my other, and equally enjoyable, research life at the University of Melbourne. As I said earlier, each institution is special in its own particular way. The University of Melbourne is a high quality research university, comparable to many similar institutions in the northern hemisphere and several in other parts of Australia. St Jude, on the other hand, is unique. This private paediatric research hospital fosters top quality research in the clinical and laboratory sciences. Desperately sick children are accepted for treatment from every state of the United States, and there have been many patients from other countries, including Russia and a number of Middle Eastern and South American nations.

St Jude is, of course, the patron saint of hopeless causes: this hospital built in his name has brought years of good life to kids who have contracted seemingly catastrophic diseases. It is also, in its small way, an agency that promotes an ecumenical view, the best of many religious beliefs and international accord. It was founded by Danny Thomas, an American of Lebanese origin and a committed Catholic, who is best known as a television actor and producer

(*Gilligan's Island, Gomer Pyle, The Andy Griffith Show*); St Jude Children's Research Hospital could have developed only in the United States of the twentieth century. Anyone doubting the generosity and kindness of ordinary Americans should take the trouble to visit it.

All children are treated free at St Jude whether or not they have insurance. One of the abiding tragedies of the United States is that some 40 million people, including large numbers of working families with small children, have no medical cover. People from places like Australia and European countries that have some form of basic, tax-payer-funded universal health care system have difficulty understanding how a country that subscribes so aggressively to a Christian perspective tolerates this situation. Part of the reason is the American passion for individual self-determination, which looks askance at the 'rationing' of health care practised in, for instance, the British National Health Service. Another is the enormous power and wealth of corporate interests, in this case the for-profit health maintenance organisations, that increasingly dominate US public perceptions.

St Jude Hospital not only provides free treatment for its young patients, but also travel and accommodation expenses for accompanying family members, counselling and psychological support for siblings, and so forth. All this costs a great deal of money. The hospital has a truly extraordinary fund-raising organisation that, in the financial year 2004, for instance, raised more that $US350 million in public subscriptions. Being awarded the Nobel gave me the opportunity to help with this by appearing on television and in other publicity formats, speaking at lunches and dinners, talking to potential donors. Needless to say, I embraced this with total enthusiasm and will continue with it as long as I can.

I've also participated in publicity and fund-raising events at other hospitals and research institutes that deal with childhood disease, including some located in Melbourne, Sydney and Perth. This does, of course, take time. When I look at the long-term future, one of my major aims is to focus on doing what I can to improve conditions for children. You might think that this is not really necessary, but much of the wealth of the world is in the hands of elderly, powerful adults. Adults vote, children don't, and poor families, in particular, often have only a minimal voice in a democracy. Sometimes it helps to remind seniors of cultural continuity, and to divert their thoughts a little from immediate needs to the question of the legacy they leave behind them.

The second commitment that claimed me was to do with the world of animal health and veterinary medicine. Shortly after the announcement of my award, Professor Charles Pilet from the venerable French School of Veterinary Medicine at Alfort, near Paris, contacted me to say that he thought I was the first person with a veterinary qualification to win a Nobel Prize. He proved to be right. Consequently, I've spoken at national and international veterinary meetings, and have also given commencement (graduation) addresses at a number of veterinary colleges. The student body is very different from my day. The sometimes rough-and-ready male-dominated classes have given way to a predominantly female, and much more stylish and sophisticated, cohort.

From the mid-1980s to 1992, I was in contact with the tropical animal health side of veterinary science through my six-year membership of the board of the International Laboratory for Research in Animal Diseases (now the International Livestock Research Institute) in Nairobi, Kenya. The mandate of ILRAD/ILRI has been to develop solutions to some of the major parasitic diseases that debilitate, and then kill, African cattle. As an old African saying has it: 'If the cattle die, the people die also'. In April 1998 I went back to Nairobi to do some publicity for the ILRI, and met a very bright young group of mainly African graduate students and postdoctoral fellows. The experience of talking to beginning scientists is the same anywhere: whether they are African, Caucasian or Asian in origin, they show the same excitement and commitment to ideas and discovery. Provide decent educational opportunities, well-funded research facilities and protection from political interference and corruption, and there is no doubt in my mind that strong science communities can be built in any country.

I was also proud to be named as an 'Ambassador for Future Harvest', a Washington-based program of the Consultative Group for International Agriculture Research (CGIAR) that administers ILRAD/ILRI and is dedicated to improving food supplies in the poor nations of the world. My fellow ambassadors include three of the Nobel Peace laureates—Jimmy Carter, Norman Borlaug and Desmond Tutu—the environmentalist and animal welfare advocate Prince Laurent of Belgium, Queen Noor of Jordan, the rock band Hootie and the Blowfish, and Muhammad Yunus, the founder and chief executive of the marvellous Grameen Bank in Bangladesh. The Grameen Bank has shown that one very effective way to alleviate poverty is to make micro-loans based on trust and with no collateral.

It has lent more than $US2 billion to 2.3 million people, many of whom are very poor women. What better evidence can there be that the best way to change things for the better is to help people to help themselves? (It would be an interesting experience if this group of ambassadors ever actually got together. This doesn't happen, but each individual's role is to help alleviate the linked problems of starvation, extreme poverty and debilitating infectious diseases. I regard writing this book as part of that ambassadorial function though, of course, real ambassadors who answer to their political masters back in their own country often have to be much more circumspect in what they say and do.)

Apart from the scientific and academic community in general, and the immunology world in particular, the other allegiance that claimed me was my original one, to my home country, Australia. I've always retained my Australian citizenship, feeling, as Danny Thomas says, 'He who denies his heritage has no heritage'. We were already committed to going to Melbourne in the November of 1996 for the wedding of our elder son, Jim, to Kate Fischer, both University of Melbourne-educated lawyers. The Prime Minister, John Howard, then recently elected, got word of this and co-hosted with the acting leader of Her Majesty's loyal opposition, Gareth Evans, a major reception at Australia's spectacular new Parliament House. An elegant public space in this magnificent building was used to entertain all of the country's senior scientists who could make it to Canberra, together with the parliamentarians who happened to be in town. It was, I think, the first time those two groups had gathered together for a celebration and it was certainly a new experience for me. Both John Howard and Gareth

Evans were very gracious and supportive of science in the remarks that they made.

Then, just before Christmas, I received a totally un-expected call telling me that I was to be named Australian of the Year. This required me to be in Melbourne for Australia Day, 26 January 1997, and to visit and speak in each of the six state capitals. Penny and I made three sep-arate trips from the United States, and met many extra-ordinary Australians in that year: people like the novelist David Malouf; the athlete Nova Peris Kneebone; the actress Ruth Cracknell; the mayors of the various cities we toured, including Jim Soorley in Brisbane and Jane Lomax-Smith (a pathologist) in Adelaide; state governors like Peter Arnison and Sir Eric Neal; and the current Governor-General Michael Jeffrey, who was then Governor of Western Australia. There were media people—the com-mentator Phillip Adams, ABC-FM's Margaret Throsby, the comedians Roy and HG—along with financial heavy-hitters like Richard and Jeannie Pratt and Lindsay and Paula Fox.

During the Australian of the Year experience, and since, I have tried hard to be positive, not to be a 'knocker'. At a business lunch in Sydney, I spoke about the lack of expertise in investment and venture capital related to science in Australia. A comment along the lines, 'It would be great if young accountants and managers could emulate the scientists and spend a year or two working in, say, bio-technology in the United States, then bring those skills back', came out in one of the major dailies with a heading like: 'Nobel Prize winner tells young scientists, go to America'. I think the subeditor wanted to run a negative story about complaining academics and simply tailored what I'd said to fit the theme. I was angry, insisted on publishing a rebuttal letter—and learned that rebuttals are a waste of time. Then there was the time I talked about the US concept of a liberal college education. The headline read something like: 'Nobel laureate suggests a society of "know-it-alls"'.

I was, I suppose, a little taken aback by the fact that the local media covered my speeches at all. The intensity and duration of media interest in the wake of winning a Nobel is probably one of the biggest, long-term surprises I've had as a laureate. I learned quickly, though, and one of the lessons is to always have some form of written account or summary to be handed around after a public lecture. Without it, your words reported by the media the next day may be unrecognisable—at least there is the risk they'll have undergone a trivialised or negative spin. Part of the inevitable and, at times, painful learning curve is realising that the only media where you can control what you say is direct-to-air radio or television.

The experiences of my time as Australian of the Year, together with my subsequent exposure to the world of public discourse, have influenced both my thought and my behaviour. Research scientists like me live much of our lives in an evidence-based world. We operate by targeting a limited question or set of questions, design experiments to test them or, in a formal sense, to disprove a 'null hypothesis' that there is no difference between our different treatment groups. This may sound a bit odd, but we often don't reach any final proof. We live constantly with uncertainty.

Scientists are not so obtuse that they fail to understand that they can't approach love, beauty or joy in quite the same way they would tackle a laboratory experiment.

Although successful scientists tend to be reasonably person-able beings who are capable of living normally in society, raising families and so forth, they do often see complex issues through the prism of evidence-based reality. In general, basic scientists are forward-looking, think in terms of underlying moral and ethical principles, tend to be mini-mally attracted by political cant and react negatively to down-market political and religious populists. Because their world is dominated by ideas that emphasise discovery, development and improvement, they are singularly un-impressed by those who look backwards and espouse reactionary and divisive views. Conversations at, say, a cock-tail party where a lot of scientists are together generally provide a robust reflection of such attitudes.

In addition, unlike politicians, scientists are allowed to hold two opposing points of view simultaneously in their heads. It's not uncommon to have lunch with a colleague who will argue some deep social issue from one point of view one day and from another the next. Such people are used to rolling complex ideas around in their heads and to looking at them from all possible angles. Those medical scientists who've trained as physicians and did internships, residencies and the like tend to be more balanced and circumspect in their public discourse because they are much more in touch with how the rest of humanity sees things.

When I was first interviewed on radio and television, I tended to speak very directly and, as I'd been accustomed to do in my normal interactions with colleagues, to go for the jugular in situations where opposing arguments seemed to be poorly thought through. After all, the people I normally deal with are at least as tough and intellectually resilient as I am. You can beat up on me all you like in the ideas sense and, like a dog chasing a ball, I will regard it just as good sport. If the discussion becomes tedious and cliché-ridden, I will, like 'Rags', drop the saliva-covered ball of an idea and trot off to investigate something that interests me. After all, at least part of the reason most of us do anything with intensity is to avoid boredom.

It can be quite counter-productive to assume that a clear, evidenced-based view of the world is generally accepted, if the aim is to get a point across to people who have different life experiences and points of view. The first lesson I had to learn is that, for the scientist at least, inter-acting with and through the media is fundamentally about communication. It's fine to be entertaining and contro-versial, but the exercise is a failure if you don't get your point across in a way that intrigues or, at least, causes people to question their assumptions.

Thinking in this way has, in turn, induced me to go back and look at many of my own long-held beliefs and prejudices, which is why, for instance, I have a chapter in this book on the interactions between science and religion. In public communications, it's worth bearing in mind the first lines of the medical Hippocratic oath: 'First, do no harm'. Nothing useful is ever achieved by humiliating people. You offend the poor individual who is crushed and alienate those who witness the performance. It's always better to lose an occasional battle to win a worthy war.

Then there is the question of interaction with poli-ticians. Politicians legitimately look to scientists for special-ist advice. According to Winston Churchill, 'experts should be always on tap and never on top'. The 'never' probably refers to that scientific adherence to evidence-based reality. Would Britain have stood as it did in World War II if

Churchill had given citizens the evidence, rather than providing the sublime political rhetoric of those great speeches: 'We shall defend our island whatever the cost may be … We shall never surrender …'? At times like this, it is an element of fantasy, the leadership of the visionary that's needed.

Politicians have both a public and a private face, depending on their personalities and the way they project their ideas and values. It is death for an Australian politician, for example, to come across as a high-powered intellectual. The former Australian Prime Minister Bob Hawke was a first class intellect and had been a Rhodes Scholar, but he hid behind a studied 'folksiness' and obsession with sport in every shape and form. As Australian argot had it, he was 'ocker Bob'. That other Rhodes Scholar, Bill Clinton, was so quick and confident that this was the political act that was expected of him. George W. Bush, on the other hand, would be in real trouble with his constituency if he suddenly started to come across as a typical product of his Yale, then Harvard, education.

Since the Nobel Prize I've had the opportunity to speak informally with a number of politicians. Whatever their public persona, many are clear thinking, committed people, who often bring that clarity of mind that is associated with a good legal training to the issue at hand. They are generally open to hearing about different ideas and ask good questions, though they won't necessarily accept what you are saying. For instance, my perception is that though Australia is yet to sign the Kyoto protocol on carbon emissions, there is a great deal of interest and concern on the part of the elected representatives themselves with the issue of global warming.

The broader Australian public is also, I think, aware of the importance of this issue and of the need for environmental protection generally. There are, of course, communities like the timber workers in Tasmania, who feel threatened by measures to promote conservation, but the overall view is that clean air and, say, the preservation of old-growth forests are good things. The level of basic education in Australia is reasonably high, but the culture is such that it is important for anyone who wants to be heard on complex issues to speak with a voice inviting democratic dialogue rather than from the perspective of the Academy or some kind of Mount Olympus. My life-long experience, from school and university vacation jobs in stores, driving vans, clearing drains, 'pulling' the rusty pipes that carry water from windmills, to being on farms and working as a veterinarian, to dealing with everyone from the lab cleaners and glassware-washers to directors of research institutes, to speaking out as a Nobel laureate on talk-back radio and in public lectures, convinces me that the majority of people are curious and that each of us forms our own particular worldview based on our education and life history.

Australians will, in the main, resist being told what to think by anyone called an expert, but they will take note of thoughtfully presented evidence-based opinion. My job through this narrative is to convey my personal understanding of science and how it works in a way that is, I hope, useful and informative. I believe it is important for the population in general, and for politicians and their advisers in particular, to approach issues like global warming with some understanding of the science informing the broad concerns that are being raised. In the end, pussyfooting around reality cannot achieve any useful result. Reality cannot be rationalised or argued away and deception can ultimately be fatal.

Politicians like to think in terms of leaving a legacy. Who would want to be seen as even partially responsible for global disaster and thus merit condemnation in the annals of history? No political leader or business executive surely would wish to attract the degree of public loathing that led Charles II to require the corpse of Oliver Cromwell to be disinterred, hanged and decapitated. Cromwell, a kill-joy and regicide, didn't do much permanent damage to Britain; his legacy in Ireland was no doubt more malevolent, but he had, like King Canute several centuries earlier, no effect on the oceans or the tides. Canute, a sensible Dane who was bored by the flattery of his courtiers, sat by the River Trent and, in vain, ordered the six-foot incoming tide to stop. Sodden, he used the experience as a lesson to his followers concerning the limitations of temporal power. Perhaps we should institute Canute Day as a UN-mandated international holiday to kick off the marketing season for the northern summer. The big stores could sell swimwear and underwater cameras, while those in the southern hemisphere could push wetsuits. Political and corporate leaders might be hosed down by their constituents or employees, and induced to reflect just a little on the warning of the historian Lord Acton: 'Power tends to corrupt and absolute power corrupts absolutely'.

Though politicians can be very tough and realistic in the course of one-on-one discussions, they can also be sensitive beings when it comes to both public exposure and the perceptions of their colleagues. This is particularly true for cabinet ministers in a parliamentary system. Votes and approval are everything in both contexts. While shining the clear, harsh light of reason and morality on some less attractive policies may cause them to wilt momentarily, it may also attract their undying enmity. Politicians are human. They may, because of the dictates of their leader or the political spectrum that they represent, be implementing viewpoints that don't, in fact, fill them with much enthusiasm. The more they dislike what they are doing, the more they will hate those who hold up the mirror.

The moral and intellectual satisfaction that might come from being an outspoken critic of this or that policy isn't worth much if it doesn't actually achieve anything. Scientists expect a prominent colleague to mediate with politicians, particularly in terms of fostering greater understanding of the importance of science and technology that will lead ultimately to increased levels of funding for them. Here, I think, the Nobel Prize and my participation on the public stage did help to achieve a doubling of both the medical (NHMRC) and the general science (ARC) research budgets in Australia, though many other individuals, organisations and two politically effective ministers of the time, namely Michael Wooldridge in Health and Brendan Nelson in Science/Education, were the prime movers.

Providing the individual is articulate, honest and interesting, the Nobel Prize does confer a long-term, public voice. What has to be avoided, though, is becoming the predictable 'Thunderer' on a spectrum of issues where there is a major, public difference of opinion. This happened to some extent to one of Australia's greatest scientists, the nuclear physicist Sir Mark Oliphant. A passionate and physically impressive man who died in 2000 at the age of 99, he was educated in Adelaide, became professor of physics at the University of Birmingham, then a research leader in the World War II Manhattan Project that produced the first nuclear bombs and contributed to ending the war in the Pacific. Though he might well have continued on to a Nobel Prize, he returned to Australia to be the founding director of the ANU Research School of Physical Sciences. I met him once in 1996, when he attended the graduation where Rolf and I were awarded honorary ANU Doctor of Science degrees; Rolf received his PhD in 1975, but it took me 21 years to get my parchment from the ANU! Chatting afterwards about the public role of scientists, Sir Mark said something like: 'They take me out of the box when they want something from me, then forget about me'. It was rather sad and a situation I certainly hope to avoid.

When it comes to speaking out there are two competing tensions. As Abraham Lincoln had it: 'Better to remain silent and be thought a fool than to speak out and remove all doubt'. Of course, Lincoln did speak out and eventually gave his life for what he believed and achieved. The other is from the eighteenth-century English statesman, Edmund Burke: 'The only thing necessary for the triumph of evil is for good men to do nothing'. If there is true evil, of course, it may take everything that a human being can offer, a prime representative of many being the Lutheran theologian and pastor, Dietrich Bonhoeffer. Bonhoeffer stood out against the nazification of his church, joined the plot against Hitler and was arrested and hanged at Flossenbürg concentration camp in April 1945 after the failed assassination attempt led by Claus von Stauffenberg. This level of commitment tends to be a one-shot deal, though as both Bonhoeffer and Nelson Mandela showed us, it can sometimes be the right way to proceed. Bonhoeffer and von Stauffenberg are martyrs of the twentieth century, but fortunately the world still has Mandela, who endured twenty-seven years of imprisonment before leading South Africa into a new era.

Most issues in a democracy, thankfully, aren't so black or white that they justify the label 'true evil'. Maybe it may seem eighty per cent bad, but the other twenty per cent can sometimes surprise. Like many scientists, I was never a great enthusiast for US president Ronald Reagan, though most accept that his actions and aggressive military spending contributed substantially to the collapse of the Soviet Union and the liberation of millions of people, including some colleagues from Eastern Europe and Russia who are spectacular research biologists.

Taking a very critical public posture is often the rule rather than the exception for those who win Nobel Literature or Peace Prizes. The Peace laureates will almost inevitably be public figures who may or may not be popular with their own and other governments. When you think about it, only a wealthy philanthropist can function as a private humanitarian. Such people can make an enormous contribution, but those who choose to remain anonymous clearly believe that virtue is its own reward. Individuals like the international anti-landmines campaigner Jody Williams (Peace, 1997) or the anti-nuclear advocate Linus Pauling (Peace, 1962) were certainly not putting out a message that was popular with the military/industrial complex (as President Eisenhower had it) of their day. Aung San Suu Kyi (Peace, 1991) remains imprisoned in Burma, and the advocate for women's and children's rights, lawyer Shirin Ebadi (Peace, 2003), seems to be under constant pressure in Iran.

The Literature laureates are expected to be direct in their comments about social and political issues, and it's generally accepted that they can be a bit grumpy at times. Those who do not write in English may be already prominent local figures before the Nobel, while people like V. S. Naipaul (2001) and Saul Bellow (1976) were already well known on the international scene. Some who write in another language, like Günther Grass (1999), will have been widely read in translation. Great literature is characterised by a critical voice that gives new insights into who we are and how we function in the world. Like science at its best, it explores basic truths. I doubt that any politician would ever expect to read a positive statement about him- or herself from a major literary figure, and the popular culture in general may also come in for a heavy caning. Other literature Laureates, like the Irish poet Seamus Heaney (1995), speak in a sympathetic voice that goes to the heart of the human experience and the society that forms them, and are loved rather than treated with a mixture of respect and apprehension.

Australia has produced one Nobel Literature Prize winner, Patrick White (1972), and is now the home base of another, the South African novelist J. M. Coetzee (2003). His recent book, *Elizabeth Costello*, has an Australian novelist as its central, rather difficult, character. Patrick White was also known both for creating some manipulative females and for being extremely difficult himself. A recent exhibition at Australia's National Portrait Gallery, housed in Canberra's original Edwardian-era Parliament House (not the pancake), displayed material on all the Nobel Prize winners who have some major, direct connection with Australia. The marvellous Brett Whiteley portrait of Patrick White, who died in 1990, displays an easily readable list of White's pet loves and hates in one corner. I found that I shared many of them. If he were alive and voiced those 'hates' today, the media would be accepting and dismissive: 'That's just Patrick White sounding off again'. If I did the same thing, hot coals would be heaped on my head: I would be stepping outside my accepted role. Of course, I could change that if I worked at it, but what would be gained?

For the moment, at least, I remain focused on the idea that achieving a good outcome can be more important than just scoring a point. Though not infallible, diplomacy, discussion, persuasion and compromise can be much more effective than going with every gun blazing. This is particularly true if the person at the forefront of the battle is poorly informed, which is often the case if you're brought in late as some sort of high-profile 'public defender'.

I have a strong personal sense of being an intellectual child of the European Enlightenment and the age of reason, holding values that increasingly need to be defended in the contemporary world. It's ironic that the United States, with its extraordinary Declaration of Independence, Constitution and Bill of Rights so grounded in the enlightenment ideas embraced by Thomas Jefferson, James Madison and their colleagues, should be the place where this attack seems most dangerous. The expectation that a Nobel Prize winner can speak with authority on almost any broad issue is, of course, absurd. There are pressures to do this as contemporary Western societies suffer, in general, from a lack of public intellectuals. Of course, the media will derive as much satisfaction from tearing down the pretensions of such people as they will from the initial dissemination of their viewpoints. A public voice should always be used judiciously.

Committing words to paper provides a discipline that forces me to look more critically at my underlying assumptions. In the following chapters I discuss some of the issues that have come to preoccupy me. As a scientist, I've seen too many 'intuitively obvious' conclusions overturned by new evidence to be able to accept even the suggestion of infallibility in anything. Scientists aren't gods or even popes, though a few deluded souls may see themselves as cardinals. It doesn't hurt to retain a sense of perspective and your own absurdity on those inevitable occasions when, as the first President Bush so concisely summarised it, you step in deep doo-doo. It helps to have a sense of humour and, when you're talking the talk and walking the walk, to look down as well as up.

6

The Next
American Century?

The types of science that are recognised by Nobel Prizes deal with universals that recognise no national or international boundaries. Both the contributions to human knowledge and the resulting technologies are potentially available to all. But the practice of science, its funding and the regard in which it is held differs from one society to another. These differences can influence the careers of individuals and the fate of nations, and can also have profound effects on humanity as a whole, and the survival of our species.

Collectively, we have made huge progress: the population of the world has burgeoned and more of us live good quality lives. In some crucial aspects, however, the way science is organised now reflects the old world. Modern research scientists are a bit like the masons of mediaeval Europe, who moved and worked among cities and states, leaving as their legacies the great cathedrals we marvel at today. Energetic, imaginative people have always migrated to where major resources are being used to create something magnificent, no matter what the tools and materials are —stone, glass, integrated circuitry or molecular genetics. Creative scientists are drawn to well-resourced, supportive centres and to enthusiastic colleagues. In this new century, though, any nation that wants to be economically strong and independent needs to attract and retain such people. The emphasis has to be on openness, education, innovation and knowledge. For those aiming to develop a culture of discovery, novel solutions—and incidentally, winning Nobel Prizes in the sciences—the central question is: How is it best to do this?

For the past fifty years, the United States has been a model of this kind of culture—the entire twentieth century was, in fact, widely regarded as the American century. Relatively open immigration, the broad values of hard work and individual commitment, along with innovation and the aggressive commercialisation of novel technologies have all delivered enormous economic success and power. Underpinning that has been the development of a high quality university sector and the extremely generous federal funding of basic research. The effect is obvious in the distribution of Nobel awards for the United States, Germany, Britain and France. Less than 30 per cent of the Nobel Prizes for science went to US residents in the first half of the century, compared with more than 70 per cent in the second half. The comparable rankings for Germany, Britain and France were 30 and 10, 15 and 11, and 9 and 3. The relative levels of research support have obviously been a significant factor, though the effects of occupation and the Nazi era also played a major part in continental Europe.

We may ask whether the twenty-first century will also be the American century. Other nations have learned from the US experience and are ploughing money into research and building their science base. As standards of living improve globally with the international dissemination of manufacturing activities, the lifeblood of the most prosperous economies will be the insight and inventiveness of imaginative, talented people. Even countries like the United States, justly proud of its profile as a centre of knowledge generation and innovation, must pay attention to this. Human capital is the most important capital. It's all very well to be bankers and accountants to the world, but that was once the role of city-states like Venice and Florence, which now figure essentially as archives of medi- aeval art and tourist centres. Lose the drivers of discovery and technology development and the financial sectors are likely to follow.

There are now many highly competitive research operations in both the physical and biological sciences throughout Europe, as evidenced by the Nobel Prizes for science over the past twenty years, but two factors have been working against European science and they both involve keeping good people. One is that it can still be more difficult than in the United States for a beginning research investigator to establish a fully independent career. The other problem is that some European countries have excessive regulation and a cultural hostility to the types of genetic engineering approaches that are at the heart of modern biotechnology and biomedical research.

Then there is Asia and the Pacific region. Every Asian country I've visited over the past ten years is emphasising the central role of science and technology for the future. Singapore has built the Biopolis, a magnificently resourced institute for molecular sciences. The president of the Taiwan Academy of Sciences, Yuan Tseh Lee, who won the Nobel Chemistry Prize in 1986, has played a pro- minent role in the spectacular expansion of Taiwan's re- search and technology base. In general, Asian politicians see the capacity of their nationals to win Nobel Prizes in the sciences as an indication they are succeeding in establishing a strong scientific culture. Japan takes pride in the fact that twelve Japanese have been awarded Nobel Prizes, but is ecstatic because three who actually work and live in Japan have been recognised over the past five years. These people are treated as celebrities (which must be tiring for the two who are now in their eighth decade).

Asian countries, however, continue to lose many of their best and brightest to the West. Over the sixteen years or so I've been associated with St Jude Children's Research Hospital in Memphis, we've seen the huge increase in numbers of young graduate students and post-doctoral fellows from Asian countries, particularly China and Korea. Many want to stay, but it looks as if the recruitment process has slowed since more stringent travel and visa restrictions were introduced following the terrorist attacks of September 11, 2001. This may adjust naturally, but if it doesn't, the United States could find it difficult to maintain its leadership of the sciences in a competitive global scene. Bright young Asians may go elsewhere.

Australia has also benefited from the same migration of Asian talent, though on a much smaller scale. The process began somewhat earlier because of the Colombo Plan, which, like the Marshall Plan in Europe, was designed to help build modern economies in the Southeast Asia–Pacific region in the post-war period. Australia makes it relatively easy for high performers to stay in the country, and the face of academia is increasingly less European. The country, however, has its own brain drain, with almost 1 million of its 20 million or so people now living in Europe, the United States and other places.

Yet by far the most substantial research and academic culture in the South-East Asia region is still to be found in Australia where, since white settlement in 1788, exploration and development have proceeded in step with the scientific revolution in the northern hemisphere. Most early Australian scientists concentrated on the unique flora, fauna and landscape, a focus that continues with initiatives like sequencing the Tamar wallaby genome and the research that seeks to conserve the Great Barrier Reef. Since political federation in 1901, the nation has built strong universities and high-profile science in a number of areas, including agriculture and mining, radio and optical astronomy, medical research and molecular biology.

Australia has a number of world-class medical research institutes, the best known being the Walter and Eliza Hall Institute (WEHI) in Melbourne. This started as a very small operation in the early twentieth century and, with national and international funding, it has grown into an institution the equal of any of comparable size anywhere. The WEHI has continued to prosper mainly through two key factors: access to additional resources and long-term, perceptive directors who were able to set consistent policies that led to high quality outputs. The WEHI also benefits from its location: it is just across the road from the University of Melbourne, which is also a leading research institution. Many of the brightest young University of Melbourne biomedical science graduates go on to PhD training at the WEHI.

The PhD research degree was not offered by any Australian university until 1950 when, partly as a consequence of advice given by the Oxford-based Australian scientist Howard Florey who won the 1945 Medicine Prize for bringing penicillin to the world, the federal government set up the Australian National University in Canberra. The ANU was lavishly funded for its time, and comprised a number of separate Research Schools dedicated to training PhD students and to pursuing research at an internationally competitive level. Rolf Zinkernagel and I did our Nobel Prize-winning work at the John Curtin School of Medical Research, as did the neurophysiologist Sir John Eccles (Medicine, 1963).

Though the John Curtin School has produced more Nobel Prizes than any other Australian institution, it went through a difficult time during the 1980s, due to declining budgets and an organisational model that emphasised short-term directorships, equal distribution of the available dollars and rigid university tenure. This situation has now been reversed and the institution is recovering. Any interested philanthropist will be made welcome by the current director, the renal physiologist Judy Whitworth. Judy is a classical music enthusiast, an ardent sports fan and cricket 'tragic', who will rapidly dispel any stereotyped notion of serious scientists as uni-dimensional or sport-hating personalities.

There is no doubt that Australia has the matter of sports leadership well in hand, but the country now needs to look to its laurels if it is to maintain a position of local intellectual prominence in the Pacific region. The aggressive and sophisticated pursuit of molecular biology in Singapore means that Australia's place cannot be taken for granted. In addition, Malaysia and Thailand are emerging as significant centres of manufacturing, and both are building up their university systems. A number of scientists have been singing the 'wake up Australia' song, a doleful dirge that surfaces from time to time but tends to be drowned out by the waves in this relaxed and comfortable beach-oriented society.

Money is, of course, central to the development of a dynamic research culture. Nations serious about R&D aim to boost spending to at least 2.5 per cent of gross domestic product. This includes both public and private sector research, with governments using tax relief mechanisms and contracts to encourage more R&D in industry. Australia currently commits about 1.6 per cent and the United Kingdom about 1.9 per cent, with the UK aiming to increase that to 2.5 per cent by 2014. The current figure for the United States is around 2.7 per cent. These figures are not totally reliable, because different cultures vary in what they classify as science, but they do provide a reasonable basis for comparison. When it comes to translating the benefits of science for human well-being, the model for the future rests in tacit, not necessarily formal, partnerships between government and the private sector. Again, the United States is the prime contemporary example: discoveries made in university laboratories or research institutes are translated quickly to the private sector. This can happen when the scientists themselves initiate the development of a new biotechnology start-up, or when BigPharma becomes immediately involved in developing a novel discovery to the point where it may emerge as a possible therapeutic. This is exactly how the first AIDS drugs came about.

How well this works in other nations depends on two things: the level of government funding for basic science and the sophistication of investors and the business sector.

The United States provides massive federal support for biomedical research in particular. Americans focus very much on the stock market when they think in terms of wealth generation and they are accustomed to the idea that there may be some risk involved. There is a whole sub-culture of venture capitalists and technology analysts that is totally absent from simpler societies. Well-motivated individuals who have made large amounts of money are happy with the thought of supporting high-risk start-ups as 'angel investors'. From my experience, many enjoy the open interaction with scientists, who are often a totally new species for most of them. They also seem surprised by the effort and enthusiasm required, and by how little is earned by truly outstanding people compared with wages paid in their own industries, whether it's finance or the movie world.

Though the 20 million Australians think, act and live like people in the advanced world, much of the national wealth still comes from exploiting the Australian land mass (the same area as the continental United States) for tourism, mining and agriculture. Local multi-millionaires here are likely to be transport barons, property developers and builders rather than technologists and innovation-oriented business entrepreneurs. Individual investors tend to focus on housing rather than on the stock market, and vast amounts of the national treasure are tied up in retirement funds, which are not accessible for high, or even moderate, risk investment. The result is a relative absence of a dynamic, discovery-oriented business culture, compared to Western Europe and the United States. While the agriculture and mining sectors use every possible technological advance and are extremely efficient, the challenges today are to build an entrepreneurial culture focused much more on discovery. Scientists need to become savvy about the possible applications of their discoveries, and trained people and venture capitalists are needed to see potential products through to the market. Australia's own internal market is small, but we've already seen one of the few major science-oriented companies, the Commonwealth Serum Laboratories (CSL), establish a substantial presence in the northern hemisphere.

Increasingly, both federal and state governments in Australia have recognised that the lack of innovative private sector involvement is dangerous, and have tried to do something about it. One initiative that seemed to be working was massive tax rebates for high technology investment. After some abuse of the system in the beginning, the program was cut back. Another is the federally funded Co-Operative Research Centres that are designed to foster collaborations between universities and industry to bring novel products to market.

The point is nevertheless that while enlightened politicians and bureaucrats can put in place the necessary tax and investment policy reforms that promote investment and innovation, governments are by their nature unsuited to driving such activities. Dollars always help, but when those in the public sector try to initiate and direct commercial development the result is often a great deal of hype and little real achievement. Bureaucracies can facilitate, but they must also understand how to get out of the way. Unlike politics, industry has to deliver a real, not a notional, product. The bottom line is a continuing income stream, not votes on one day every three or four years.

Starting in the 1930s with the founding of the CSIRO and the later establishment of the ARC and the NHMRC, Australian governments have given reasonable, though not spectacular, levels of support for basic and applied science. The outcome to date is modestly successful. Australia has a relatively small but very high quality basic and applied research sector; and three Nobel Prizes have been won by Australian medical scientists for work done in Australia.

*C*ontributions from the public and private sectors across several countries have come together for great human benefit in the evolving story of the human papillomavirus (HPV) vaccine. A number of first class scientists, including Harald Zur Hausen at the German National Cancer Institute in Heidelberg, Margaret Stanley at Cambridge University and many others have progressively made the case that some variants of HPV cause cervical cancer in women. The idea of developing a vaccine against HPV was taken up in the early 1990s by Ian Fraser, a Scots-trained medical doctor who learned immunology while working at the WEHI, then moved to the Princess Alexandra Hospital and the University of Queensland in Brisbane.

Fraser used a novel 'replicon' strategy and, with grant support from the federal NHMRC and Queensland Cancer Society, developed his vaccine to the point where he could start interacting with CSL in Melbourne as a commercial partner. Testing this was too big a job for CSL, so the product was sold off to the US company Merck, which now has what looks to be a very promising vaccine in large-scale phase 3 trials. The likely outcome is that millions of women worldwide will benefit by being protected against HPV infection. Merck will make a lot of money,

Ian Fraser should profit too, and the Australian economy will recover the relatively small public sector investment multiplied by a factor of thousands in royalty payments. Given appropriate partnerships, this story could potentially be repeated in any developing country for, say, a new therapeutic developed from systematic analysis of the active ingredient(s) in a particular local plant product that is believed to have medicinal value.

There are positive and negative lessons from the Australian experience for countries now building a science base. One obvious one is to establish a good school and university system and to link science and business education in ways that increase the likelihood of mutual understanding and interaction. Governments can play a very positive role by taking care to establish the right tax and investment structures. It is obviously a disincentive to levy sales tax, import tax or the equivalent on scientific equipment and research reagents that have to be imported. Where resources are limited, it makes sense to focus on local problems and opportunities.

The US example provides the most successful model for distributing federal research and development funds in an advanced economy. Committees like the National Institutes of Health 'study sections', which evaluate investigator-initiated grant applications three times a year, use a rigorous peer review process that takes no account of geography or politics. If approved, the grant will generally extend over five years. A renewal of the grant will depend on what has been achieved, measured mainly by publication in prestigious, peer-reviewed research journals. It helps, of course, if the project has already led to practical outcomes and successful patents.

The grant reviewers are usually established, special-ised, mid-career researchers and scholars who give a month or more each year to study section service over four to eight years. This prevents any tendency for the funds to be con-trolled by a few powerful senior people. The US National Science Foundation (NSF) that supports research in physics, chemistry, mathematics and linguistics, among other things, operates in a similar way. In both cases, I doubt that any federal money is distributed on a better-evaluated or more ethical basis. The processes that decide on the 'big ticket' items, such as optical and space tele-scopes, however, are inevitably more political.

The net result of the peer-reviewed 'small science' mechanisms used by the NIH and the NSF is that the best and brightest young US scientists become completely independent early in their careers, and can rapidly build research groups limited only by their capacity to do innova-tive science. It isn't just the individuals who win: each grant comes with an amount for overhead expenses which goes directly to the host institution. With a typical overhead rate of 40 per cent, this means $400,000 for the university administration when an investigator receives a grant of $1 million. Not surprisingly, university deans and presidents love the competitive medical scientists who, in turn, enjoy a very active job market and relatively good incomes, as much of the salary costs are paid through the research grants.

Working in the United States, I had the sense of being a marketable commodity. Anyone seen as a potential Nobel Prize winner is hot property. As in all countries, the uni-versities vary enormously in standards. Those at the top of the pile retain their position by attracting very talented people who, incidentally, are likely to generate the most research dollars. Others down the league a little may be seeking to improve their status by recruiting a star. Academic life in the high profile institutions is a very competitive game: some who don't make it at a junior level are sent back to the minor leagues as a consequence of the up-or-out rule that applies for promotion from assistant to associate professor. This barrier can be extremely difficult to surmount in top places like Harvard, but many move to slightly less prestigious institutions and go on to enjoy substantial careers. The US system is very effective in bringing forward Nobel laureates, though the model can at times seem too goal-driven and 'busy'.

Though the university cultures are somewhat different, research in Britain and Australia has tended to operate along the lines of the US peer-review model. The total dollar amounts allocated to discovery science per head of population are smaller, and the lower indirect cost rates may not be tied so directly to the individual scientist's performance. The result is that the interaction between scientists and administrators assumes a different and more stable dynamic in the countries of the old British Commonwealth. People often do remarkably well with less funding than they would have had in a comparable US institution.

Other countries, particularly Japan and Germany, provide relatively good levels of science funding, but have suffered from a traditional institutional bias towards a much more hierarchical—and sometimes stultifying—university structure. Fairly recently, at least, a number of young Japanese and German scientists have first established their research careers in the United States and then returned at senior level to major positions at home. Leading Japanese and German academics are aware that many of them aren't comfortable with the old system after their experience in the States and are making efforts to change their university systems.

Part of the problem is that there has been a greater tendency in some countries to favour core-funded research operations. In a core-funded operation, the dollars are given to the institution and are then distributed by an all-powerful director who parcels out the resources to the individual research groups. This can work well if the director is smart, takes good advice and is prepared to be personally unpopular as a consequence of favouring the most competitive research groups. However, such decisive leadership is more often the exception than the rule.

The danger with such models of guaranteed research support are that some people become too comfortable, go to sleep and can be very hard to wake up or dislodge. The core-funding practice is pretty much restricted to independent research institutes. Unlike the situation in a teaching university where there are other jobs for staff whose interests have changed, there aren't many alternative roles in dedicated research operations for full-time scientists who have lost their edge. That's why the question of tenure in research institutes and research universities is such a difficult problem.

Among the more successful examples of a core-funded research operation is the massive NIH 'intramural' laboratory complex in Bethesda, Maryland. Scientists working in these laboratories have been awarded five Nobel Prizes. Taken together, the intramural core-funding mechanism at Bethesda and the peer-reviewed extramural NIH grants that fund biomedical scientists throughout the United

States via competitive grant applications have supported more than eighty Nobel laureates, including me. Another example of a stellar core-funded operation is the British Medical Research Council Laboratory of Molecular Biology, the iconic LMB. This is a centrally funded institute led by extraordinarily talented people who stay close to the actual practice of science. Together with its precursor, the LMB has provided the Cambridge working environment for twelve Nobel laureates. These include Francis Crick and Jim Watson (Medicine, 1962), Max Perutz (Chemistry, 1962) who founded the laboratory, and Fred Sanger, who is the only person to have won the Chemistry Prize twice, for working out how to sequence proteins in 1958 and DNA in 1980. Dollar for dollar, the LMB must be the single most successful basic biomedical research operation the world has seen. It continues to be a fine institution.

The direct support of able directors who distribute generous funds to younger, more short-term research colleagues is the mechanism used by the fifty or more core-funded institutions of the German Max Planck Society, which grew out of the pre-1945 Kaiser Wilhelm Institutes. The Max Planck Society lays claim to sixteen Nobel laureates in all areas of the sciences, while a further fourteen were recognised as members of the old Institutes. Albert Einstein (Physics, 1921) published his general theory of relativity in 1915 when he was director of the Kaiser Wilhelm Institute of Physics in Berlin. The theoretical physicist Werner Heisenberg (Physics, 1934) was appointed director of the same Institute in 1941. After the war, he moved the renamed Institute to Göttingen and led it until his retirement.

*C*an high quality basic research ever flourish in repressive, centrally organised societies? The answer is clearly 'yes' for applied military science and technology: Nazi scientists developed jet fighter planes and other effective killing machines. Apart from the fact that the Nazis drove out many of their most established scientists like Albert Einstein, and future scientists like Max Perutz, because of their Jewish heritage, the cultural values and propaganda of these ersatz Aryans made the honest pursuit of new knowledge in areas like medical genetics and even haematology impossible. They believed, for instance, that there was a difference between 'Aryan blood' and 'Jewish blood'. Worse than that, what happened under the guise of medical research in the Nazi death camps plumbed the depths to which science that is uninformed by morality, ethics and compassion can go. Science cannot be built on human degradation and lies.

Something similar happened with genetics in the USSR. Stalin embraced the arguments of the plant breeder T. D. Lysenko, which claimed that acquired characteristics are inherited, an idea attributed to the eighteenth-century French scientist, Jean-Baptiste Lamarck. It suited that totalitarian ideal that people could be 'improved' in some heritable way by social conditioning, and also suggested a rapid route to enhanced agricultural production. Lysenko was appointed director of the Institute of Genetics in the USSR Academy of Sciences, while his predecessor, N. I. Vavilov, was exiled to Siberia. This effectively destroyed research in genetics for twenty years and contributed to the later inefficiencies in food production that proved a major embarrassment for Russian communism.

The only Russians ever to win Nobel Prizes for Medicine are Ivan Pavlov, of the Military Medical Academy of St Petersburg in 1904, and Ilya Mechnikov, who worked at the Institut Pasteur in Paris, in 1908. Pavlov survived the 1918 revolution and, though he was for a time highly critical of both the communist regime and Stalin, lived to the age of 87 and died of natural causes. His experiments on conditioned behaviour in dogs evidently appealed to the 'man of steel'. Though some of the techniques and ideas that Pavlov developed were later used in the treatment of psychologically disturbed humans, he himself was a decent and courageous man and can in no sense be blamed for the notorious abuses of psychiatry that were to develop in the Soviet system.

Stalin died in 1953 and the USSR began to slowly change. Since 1958, ten Russians have been awarded the Nobel Prize in physics, including Alexander Prokhorov (1964), who was born in 1916 on the Atherton Tableland in northern Australia but moved to Russia with his parents in 1923, after the revolution. The physical sciences and their associated technologies were obviously facilitated in communist Russia, as evidenced by the fact that in 1961 they beat the Americans into space. This caused major concern and had, in turn, an enormously positive effect on science funding in the United States. President Kennedy promptly stated his goal of putting a man on the moon.

In the areas I know about personally, Russian medical scientists did a good job in developing influenza vaccines and in working out the epidemiology of the tick-borne infections that cause problems in spring and summer. Of course, they also developed a sophisticated capacity to make bio-terror weapons. Although these programs have been closed down, there is still concern about the fate of Russian stocks of lethal, genetically modified smallpox viruses. All this type of work is now focused on protecting human populations, not the opposite, and the Biological Weapons Convention of 2002, endorsed by 147 countries, was created to ensure that this situation continues. As with all leading nations, the United States has long abandoned biological warfare as a strategy and turned the supporting military facilities, like the Fort Detrick complex in Frederick, Maryland, over to the study of dangerous, so-called emerging, diseases like Marburg and Ebola virus infections. These bio-security laboratories are also being used to develop strategies and reagents for countering possible bio-terrorism; but the perception is that any such attack would come from groups of isolated zealots rather than from the agencies of an established government or nation-state. Overall, biological agents are not perceived as being particularly efficient military weapons but, as the anthrax episode in the United States that followed shortly after the September 11 catastrophe showed, they are effective at promoting disruption (of the postal service in that case) and general fear.

The national scientific dynamic of any country, whether or not it has a substantial university and science sector, can gain a measure of diversity with the involvement of international partners. It is difficult to build a science-based culture in pastoral or agricultural societies where family and tribal allegiances and traditional modes and values prevail. There is also the 'big man' phenomenon that sees scarce dollars rerouted to secret Swiss bank accounts and so frustrates Aid agencies. Cultural models that rely on networks of family or tribal interactions have to be broken down or marginalised if true research excellence is to emerge. One way of getting around this is to develop inter-actions with powerful international agencies, to produce the kind of global collaboration that might defeat tuber-culosis or malaria. This is the focus of the Bill and Melinda Gates Foundation's 'Grand Challenge' grants, which link multi-institutional, multi-national research operations in the advanced world to target these major infectious disease problems in the developing countries.

Agreements between the Institut Pasteur in Paris and the Chinese and Korean governments have created a unique model for interfacing a not-for-profit private foundation based in the northern hemisphere with newly developed state-funded enterprises located in Asia. The Institut Pasteur de Shanghai/Chinese Academy of Science was officially opened in October 2004 with a French director-general and a Chinese co-director. The Institut Pasteur of Korea in Seoul also began in the same year with a French director. In both cases, the financial underpinnings are pro-vided by the host nation. The original Institut Pasteur in Paris began establishing regional off-shoots in 1891. There are now twenty of them, including those established in Benin in 1961 and Ivory Coast in 1972.

The institute's founding father, the great micro-biologist and humanitarian Louis Pasteur, died in 1895, the year before Alfred Nobel. Since Mechnikov in 1908, seven other Paris-based Pasteurians have won the Medi-cine Prize, including Francis Jacob, Andre Lwoff and Jacques Monod, who shared the 1965 award and who are key figures in the founding of the science that is changing our world, molecular biology. At the Pasteur Museum in Paris, now surrounded by modern research laboratories, one can see the fully furnished private rooms where Pasteur lived on one side of a corridor, his laboratories on the other. The contemporary Institut Pasteur is funded by a variety of mechanisms, including approximately 50 per cent from French government resources via a spectrum of peer-reviewed and institutional support processes.

Based in Lyon, the for-profit Institut Merieux did extremely valuable research at the more practical end of the spectrum. This 'translational' aspect of science is enormously important and is sometimes recognised by Nobel Prizes: the 1991 Medicine Prize, for example, went to two Americans, Joe Murray and Don Thomas, for pioneering the techniques of organ and bone marrow transplantation. The only Nobel Prize for Medicine ever given directly for the development of a viral vaccine went to Max Theiler of the Rockefeller Institute of Medical Research—now the Rockefeller University—in 1951 for the 17D yellow fever vaccine that is still in use today.

The Institut Merieux is now part of the world's largest vaccine manufacturer, the commercial Aventis Pasteur operation, which also incorporates the north American Connaught vaccine development and production facilities. Over the past ten years, amalgamations have been the rule in what is known as BigPharma. This extensive restructuring of the drug industry, with the reduction in staff numbers that results inevitably from mergers, has changed some of the realities for scientists working in areas like pharmacology. Some big companies have essentially dismantled a substantial proportion of the 'R' component of their R&D operations, and are now relying a great deal on the innovation and discovery that comes from the public sector, or from small high-technology start-ups that they then absorb.

In the past, however, it has certainly been the case that Nobel Prize-winning discoveries have emerged from such essentially profit-driven, private research centres. John Vane of the Wellcome Research laboratories in Beckenham, UK, shared the 1982 Medicine Prize with the Swedish scientists Bengt Samuellson and Suni Bergstrom for their work on prostaglandins. Wellcome scientists Gertrude Elion and George Hitchings from the Research Triangle in North Carolina were recognised by the 1988 Medicine Prize, along with James Black of King's College Hospital Medical School in London, for illuminating important principles in drug treatment. Other Nobels were awarded for discoveries made by physical scientists working in private sector institutions like the US Bell Laboratories. These prizes, which recognised many of the advances in electronics, computing and communications that have so changed our world, were often shared with researchers from leading universities.

The Bell Laboratories that were funded directly by the Bell Telephone Company gave scientists enormous freedom —freedom from both having to raise money and in the research directions they could pursue. A similar situation applied at the Basle Institute for Immunology, an independent, basic science laboratory that was fully funded by the Swiss pharmaceutical company Hoffman La Roche. In the thirty years of its operation, the Basle Institute provided the working environment for a spectrum of outstanding scientists, many of whom are now in the United

States. It has been sorely missed since its closure in 2000, but it was extraordinary that Hoffman La Roche provided such generous, no-strings support for so long.

The Basle Institute was unique in that, apart from the position of the founding director Nils Jerne, who had turned his mind to theory and no longer led an active laboratory program, there was no hierarchical structure. All the scientists, whatever their age or eminence, were identified by the title 'member' and were theoretically equivalent in stature. I like this idea a great deal but, in practice, although all the members were equal, some (like the animals in Orwell's *Animal Farm*) were inevitably more equal than others.

A geneticist from the institute, Susumu Tonegawa, won the 1987 Medicine Prize, and Nils Jerne, together with two other scientists, won the 1984 Medicine Prize. Jerne made important early contributions to immunology, as described in chapter 4, and late in his career he developed the still controversial 'network theory' of immune regulation.

The Wellcome Laboratories have also been transformed by sequential amalgamations and are now part of the Glaxo Smith Kline company. What survives magnificently is the Wellcome Trust, established in 1936 following the death of the company's founder, Sir Henry Wellcome. The available resources were immensely boosted when the Trust sold off its shares in the former Burroughs Wellcome drug company. The Trust is now the largest charity in Europe, using a combination of peer-reviewed grants and fellowships, together with specific allocations for large projects like the Wellcome Trust Sanger Centre for genome research, to distribute in excess of a billion pounds sterling annually. The much smaller Burroughs Wellcome fund, based in North Carolina, distributes more than $US25 million annually for biomedical research in the United States and Canada.

The other great example of a research support organisation developed from the bequest of an industrialist is the Howard Hughes Medical Institute, funded from the estate of the eccentric and reclusive billionaire. Operating as a 'virtual' institute, it pumps more than $500 million a year into basic biomedical research. The administrative headquarters are in Maryland, and its scientists are scattered through the country's leading universities and research institutes. Though most Hughes funding goes to US residents, they have also provided some support for international programs in infectious disease and cancer led by scientists in Australia, Eastern Europe and so forth.

The Hughes Institute is an unashamedly elitist organisation. The central idea is that the very best scientists should be free to concentrate on their research, and should not have to be constantly writing research grant applications. Leading scientists and younger research investigators who show evidence of exceptional promise are invited to join, and are then provided with very substantial funds. After they have come through the up-or-out process, their performance is reviewed at regular intervals. To date, thirteen Howard Hughes scientists have been awarded Nobel Prizes, including Linda Buck and Richard Axel who shared the 2004 Medicine award for working out the neurological basis of olfaction, the sense of smell.

The most recent US Nobel Prizes for research done in the for-profit commercial sector went to William Knowles of the Monsanto Chemical Company for Chemistry in

2001 and Jack Philby of Texas Instruments for Physics in 2000. Both had retired by the end of the 1980s. The latest award to a scientist still working in the chemistry/pharmacology industry was the 2002 Chemistry Prize to Koichi Tanaka of the Shimadzu Corporation in Japan. This could be a sign for the future. The large US drug companies are, for the reasons I discussed earlier, much less likely to be supporting innovative, in-house research. By concentrating on the development part of R&D (clinical trials and so forth), they are in effect reacting to the reality that, with the application of contemporary molecular approaches, major new discoveries can emerge anywhere—and sometimes from the most unexpected places. The involvement of BigPharma and the profit motive is, however, absolutely essential if any such findings are to be exploited for eventual human benefit.

Since the 1970s, one of the major areas of excitement has been the new biotechnology industry. It blossomed in the wake of Watson and Crick's 1953 model of the DNA double helix and progressed through the development of restriction enzymes to cut DNA to the development of recombinant DNA technology. The scientific founders were generally university scientists, though those who succeeded also had the good sense to involve leaders with business skills from the outset. Some of the early players, like Genentech, DNAX and Immunex, have been tremendously innovative and it would not be surprising to see Nobel Prizes going to researchers who did their best work in such operations.

Every city from Cambridge, Massachusetts, to Brisbane, to New Delhi, to Cambridge, UK, that hosts major research universities now has an associated 'biotechnology cluster' that can number hundreds of small start-ups and larger companies. Many of these operations develop their intellectual property to the point that it is either attractive for the drug companies to buy them *in toto* or to purchase the technology they have developed. A number of scientists who began as modestly paid but intellectually driven research workers have become wealthy through this process, proving that a Nobel Prize is neither the sole nor the most lucrative reward for a research scientist.

The small biotechnology operations that take one or a few discoveries part-way down the track to commercialisation have emerged naturally in those nations that have a major publicly funded science base. Many smaller countries are also hoping to stimulate economic growth by directing their more limited resources towards biotechnology. In Cuba, for instance, such developments are being driven from the public sector to provide opportunities for bright young scientists.

As with the wealth of nations, there tends to be a north–south divide in science. Very expensive facilities, like the billion dollar accelerators used in high-energy physics, are generally found in the prosperous countries of the northern hemisphere. For instance, the Tesla accelerator that is currently being constructed in Hamburg is costed at 684 million euros. The exception is the Brazil synchrotron, a smaller scale accelerator that, among other applications, is used by the contemporary equivalents of Max Perutz and Rosalind Franklin to determine protein structure. Australia has started to build a synchrotron, but for now Australian structural biologists see themselves as 'suitcase scientists' who travel to Narita, Hamburg or Chicago to do their key experiments. Apart from the inconvenience and loss of time, transporting even the types of non-living materials they study has become more cumbersome with the increasingly stringent regulatory requirements spurred by concerns about bio-terrorism.

Obvious exceptions to the north–south rule are the big optical telescopes, where the ideal site may be in the high, clear air of the South American Andes. The twin 8-metre telescopes of the Gemini Observatory are located on Manua Kea in Hawaii and on Cerro Pachon in Chile. These instruments are supported by international partnerships involving many nations. Some of the Associated Universities for Research in Astronomy telescopes are to be found near the city of La Serena in Chile, which seems a very appropriate place, as optical astronomers tend to be calm types who enjoy sitting on top of mountains. The southern sky has obviously been the focus of the various Australian observatories, with radio and optical astronomy traditionally being very strong on the Australian science scene.

International consortia are, of course, enormously important when it comes to seeking scientific solutions to the infectious diseases, like malaria, which continue to kill so many in the poorer countries. The efforts to deal with the AIDS catastrophe, where some 3 million people, including 1.2 million women and 600,000 children, died in 2002 alone, constitute a massive global enterprise. Candidate AIDS vaccines must ultimately be tested in high-incidence areas, so we see the development of interactive networks that bring together scientists, ethicists, health care delivery personnel, administrators and a new breed of specialists who interact with regulatory authorities. A potential vaccine developed in the United States will have to be approved by the FDA and by the comparable regulatory body in the country where it is to be tested. The design of the trial will be subject to the requirements of the human subject experimentation committee at the institution where the product is developed and by the authorities in the recipient nation.

In developing countries, where catastrophic infectious diseases like tuberculosis and malaria are still rampant, international collaborations and funding are having a dual outcome. A major advantage of involving both altruistic foundations and government-supported Aid agencies is that they can sometimes set up substantial research operations that train young local scientists and provide opportunities for them. As I mentioned earlier, I was for six years a board member of the internationally resourced and directed ILRAD/ILRI research institute in Nairobi that had the job of trying to deal with African trypanosomiasis (sleeping sickness) and theileriosis, a fascinating disease caused by a malaria-like parasite that infects white rather than red blood cells and makes them essentially cancerous. In the process of trying to solve these very difficult disease problems with relatively limited (compared to TB and malaria) resources, ILRAD/ILRI trained a number of first-class young African molecular biologists and epidemiologists who now work in all types of related areas.

Generally, however, there is no global, level playing field for young scientists who want to seize available opportunities and develop their own potential. The answers to the important questions—Where can I best train and where should I ultimately aim to develop my own independent research program?—are not at all simple for those born outside the large open economies of the United States or Europe. Many scientists from the developing world who work for a time in the universities and research institutes of the advanced countries find it difficult to return home. The resources to pursue their particular passion may simply be unavailable.

However, both the laboratory and funding requirements for molecular biology and biotechnology are relatively modest and, as these approaches have the potential to help solve issues relating to food production and human and animal health, they are of particular value for developing countries. If these societies have a vested interest in high technology products and approaches, communities are more likely to accept them. What more appropriate target for Aid dollars and international collaborative arrangements could there be than building this type of infrastructure, with the consequent growth in local expertise and national pride?

Science at its best is a universal activity that benefits everyone. Positive international change is more likely to be achieved if scientific discoveries and advances are seen as broadly 'owned' by all members of the human family. Ownership implies a measure of understanding and the sense that people are involved in, rather than being exploited by, the linked processes of discovery and development. Education alone is not enough. The need is to build local, scientifically aware cultures that not only provide jobs for bright young people but also influence the ways that individual citizens, entrepreneurs and governments approach the challenges and the opportunities that are open to them. To my mind, 'challenge' and 'opportunity' are two faces of the same coin.

There is no simple answer to the global inequities in wealth and the spread of resources for national research enterprises. Nevertheless, approaching science in a targeted and thoughtful way that emphasises local, selective advantage and altruistic international partnerships has real potential for building one or more areas of national research excellence. A major breakthrough with malaria, AIDS or starvation could well result in a Nobel Medicine or Peace Prize to a scientist working in a developing nation. I can't think of anything that would be more likely to delight the Swedish and Norwegian Nobel committees than being able to make such an award, especially in the sciences.

Meanwhile, advanced nations like the United States and Australia need to do more to keep ahead of the game. The United States is currently suffering from a rise in anti-science and anti-intellectual attitudes fuelled by parochialism and political and media pandering to narrow religious fundamentalism and the perceived interests of large corporations. However, the United States swings dramatically in broad public attitudes and practices, and this will hopefully be only a short-term trend. Any student of the United States is acutely aware of how extraordinarily successful the country has been in the process of periodically re-inventing itself.

Australia, with its small population and its South Pacific location, needs to use the years of respite provided by its vast natural resources to create a greatly enhanced, high technology industrial base. When it comes to the harsh realities of the mining and agriculture sectors, Australia has fulfilled the prediction of Arnold Toynbee's theory of history that the creative minds that drive innovation come to the fore under conditions of crisis and extreme difficulty. The penalty has been that resultant prosperity has been sufficient to marginalise the sense that the nation must build a much broader, and more dynamic, culture of innovation and development if it is to remain prosperous and be a significant international player in the long term. This book is aimed at generating greater awareness around this issue. As Prime Minister John Howard said to me: if you want to influence the political process, get the voting public interested and involved.

7

Through Different Prisms: Science and Religion

Is a faith-based view of the world in irrevocable conflict with science and with developing new knowledge? Does adherence to religious tradition and practice limit the willingness of a nation or an individual to embrace new ideas and ways of looking at the world? Is the tolerance of social diversity that accompanies a dynamic science culture a threat to religion? Can scientific discovery and theory be reconciled with religious belief, or is useful dialogue between these value systems impossible?

Science is an activity that suits people who question, test ideas and then embrace the intellectual and philosophical consequences of their findings. There are no absolute or revealed truths in science: any belief or theory will soon be discarded and forgotten if it no longer fits the available evidence. The differences in scientific and religious cultures mean that people who hold to literal, faith-based views can find this difficult to understand. The theory of evolution is attacked because 'it's only a theory', but the fact is that evolutionary theory has been the single most important explanatory model in biology for more than a century and it underpins many of the key breakthroughs in medical science recognised by Nobel Prizes over the past fifty years.

In the media, the public debate between science and religion is often posed at the extremes, creating a superficial perception that adherence to the one means automatic hostility to the values and practices of the other. In fact, though there are obviously situations where no intellectual compromise can be reached, this rarely spills over in a way that limits the useful application of science for the general good. These two cultures can and generally do live side by side in reasonable harmony and, given the massive problems that humanity will face over the next century, they must be able to talk to each other.

To borrow from St John's gospel, I would say that science is a 'big church' with 'many mansions': it accommodates all sorts of people raised in many different belief systems. My own background is that I grew up in a substantially secular household that had a strong dedication to reading and education. Both my parents had been to the local Methodist church as children and, thinking about it now, it seems to me that my mother sublimated her spiritual life into the piano and her roses, while my father was heavily involved in the Masonic Lodge. My paternal grandmother was a committed Methodist, who derived great solace from the church after the early death of her husband. He was a lapsed Catholic, who continued to read and mark passages in the Bible after leaving the church. The family story has it that the breach occurred when he returned after a long absence to his family home near Ballarat, Victoria, and found that the local priest had burned some of his books because they were proscribed. My maternal grandmother, on the other hand, was brought up in a Quaker household and went to the famous Quaker boarding school at Saffron Walden in Essex. By the time I knew her she had totally rejected religion and devoted her life to her family, her chickens and her vegetable garden.

I attended the Methodist Sunday School where the annual book prizes for attendance, which was a bit spotty in my case, were the only ones that I ever received before I started winning international prizes for medical research. Like many teenagers of that era, I went through an intense religious period at about age 15 or 16, spent time at evangelical church camps and heard Billy Graham preach during his famous 1959 crusade to Australia. Since then, I have avoided any situation where crowd dynamics are likely to dominate.

Adolescence is a period of conflicting influences. At this time I was also going to army cadet camps, where I learned to fire the lethal Lee Enfield rifle. I vividly recall using an Owen sub-machine gun to blow holes in 'human' targets that popped out from behind trees at the Defence Force's Canungra jungle training centre. Even for 15-year-olds, the training was highly professional. There is nothing romantic about guns; they are not toys and the primary purpose of a military weapon is to intimidate, disable or kill. Nothing in my experience led to the confusion of God and guns that figures so prominently in some extremist agendas. Perhaps predictably, neither the gun phase nor the religious period survived my early years at university. Though that has been my choice, I respect those who pursue the profession of arms in the defence of peace and stability, or dedicate their lives to the service of others from a sense of religious vocation.

University was my first introduction to real people who live by ideas and experiments, though I already had some insights into that world from reading. These years also began my life-long intellectual adventure with biology, and biology makes no sense whatsoever in the absence of

Darwinian evolution. This is particularly obvious for the late-evolving adaptive immune system that I have described earlier, which operates more like a street person dressed in cobbled-together hand-me-downs (the molecular mechanisms used by 'older' systems like the brain) than a perfectly arrayed socialite in an elegantly accessorised Armani or Versace outfit. Immunity does not look to me like something that is 'intelligently designed'.

Once I had made the transition in thinking to evidence-based enquiry, there was no going back to assertion and dogma. The consequence is being eternally condemned to an intellectual view dominated by verifiable data, reason and, so far as any human being can achieve that, self-knowledge. That doesn't mean that I've set aside early positive influences that came from the non-conformist Protestant tradition. I also realise that, to some religious people, my reality-oriented view of the world must seem like a vision of hell.

My childhood left me with a reasonably good, if superficial, knowledge of the four gospels in the magnificent King James Bible. I dip into them from time to time, and tend to quote religious texts, probably incorrectly, at some of my more religious friends in the American south. To me, the Bible represents the collective wisdom and stories of the Jewish and early Christian people. The oldest of the gospels, St Mark's, was not written until forty years after what is generally accepted as the time of Christ's death. I am clearly more impressed with statements like 'It is easier for a camel to go through the eye of a needle than for a rich man to enter the kingdom of God' (Matthew 19:24), 'Sell whatsoever thou hast and give to the poor' (Mark 10:24), or 'Judge not, that ye be not judged' (Matthew 7:1) than many religious Americans. Like everyone else who has some religious background, I quote selectively from the bits that suit my particular experience and consequent view of life and the world. Many of us, when it comes to thinking about religion, relate to an early childlike experience of church-going that is fairly narrow and unsophisticated, and in total ignorance of the vast mass of religious scholarship and theology.

The Methodists also have a great dedication to hymn singing, which I sometimes inflict on those who are unfortunate enough to be in my immediate vicinity at the time. Entering the great European cathedrals leaves me with a profound sense of immanence and inner peace, which is exactly what they were designed to do. One of my most vivid memories is of being at the deathbed of my religious grandmother as a 10-year-old child. While she was dying, she recited the 23rd Psalm, 'The Lord is my Shepherd; I shall not want …' This calming mantra goes through my head when I am deeply troubled about something and can't sleep. Saying 'om' over and over does nothing for me and I don't count sheep, but I have an olivewood carving of the Good Shepherd carrying a lamb that I bought in a Bethlehem tourist gift shop many years ago. The shepherd, with his humble status and simple dedication to protecting the vulnerable, seems to me to symbolise the very best of the Christian tradition. The Bethlehem shepherd stands on my bedside table.

The point I would make here is that for anyone thinking about a career as a research scientist, a religious upbringing should not be a problem. The father of the 1988 physics laureate, Jack Steinberger, was a cantor and religious teacher, while that of the 1958 winner in Medicine,

Josh Lederberg, was a rabbi. Mike Bishop (Medicine, 1986), Ed Krebs (Medicine, 1992), Bob Curl (Chemistry, 1996) and John Sulston (Medicine, 2002) grew up in households headed by Lutheran, Presbyterian, Methodist and Anglican ministers respectively. The difficulty is for those from fundamentalist backgrounds that emphasise the literal truth of holy writ. Can such a person hold on to that faith on the one hand while on the other pursuing a scientific career based in modern molecular medicine? At some stage, the individual concerned will be forced to confront, and deal intellectually with, the mass of evidence for evolution and natural selection.

The need to discuss religion in a book about science would have been considerably less obvious twenty or thirty years ago. Much of Western Europe had moved towards what is now often described as a post-Christian civilisation, with even those who had some religious affiliation practising 'cafeteria Catholicism' and the like. The German philosopher Friedrich Nietzsche had long since declared 'God is Dead', a view that resonated in the influential book *Honest to God* by the Anglican bishop John Robinson, published in 1963. The fact that many US intellectuals were in denial about the lives of those who inhabit 'middle America' had pushed any consideration of religious fundamentalism into the deep background. After Sinclair Lewis, the 1930 Nobel Literature Prize winner, created the despicable evangelist Elmer Gantry, I can't think of a major US novelist who has addressed the outer limits of Christian practices in the American heartland. Maybe it would have been impossible to invent characters who were more extraordinary than Jimmy Swaggert, or Tammy Fae and Jim Bakker.

Beginning in the 1990s, however, substantial new books that look analytically at religious fundamentalism have been emerging with increasing frequency. The most helpful account that I've read is Karen Armstrong's *The Battle for God*, which provides thoughtful, and even sympathetic, insights into the extremes of the three Abrahamic religions, Judaism, Christianity and Islam. John Krakauer's *Under the Banner of Heaven: A Story of Violent Faith* is particularly intriguing because it relates the very recent and well-documented origins of Mormonism, while at the same time focusing on the more regressive fringes of that belief system. Krakauer also documents the way in which American fundamentalism can be tied to the libertarian philosophy which holds that the individual has no social obligation to pay taxes or to obtain a driver's licence. Combining the fundamentalist and libertarian themes with the US gun culture gives the lethal mix that led to the 1993 tragedy in Waco, Texas, where seventy-six members of an extreme cult died in a confrontation with the secular authorities.

The reality that religious fundamentalism is a major political force was made clear during the 2004 US presidential election, and in a way that was almost incomprehensible to observers elsewhere. The United States is the most conventionally religious country in what we traditionally think of as the Judaeo–Christian world. The presidency of George W. Bush has left us in no doubt that a substantial percentage of US citizens think of themselves as 'born again' Christians.

The big area of conflict between science and religion is, of course, to do with the belief system known as 'creation science'. Some religious people find the idea that humans have evolved from simpler life forms to be unacceptable. Others go further and accept that every word in their particular translation of the Bible is literally true, which means that, when the dates are calculated, the world can be only about 6,000 years old. At this extreme, the creationist argument has to rationalise the biological and geological record as either some sort of theistic trick, or a test of faith. At the other end of the spectrum is the idea of 'intelligent design', which accepts evolution but argues that the process is not random, but God-directed. Biologists and 'intelligent design' people can at least talk to each other, though I doubt they will ever agree.

Why should science be unduly bothered by creationist arguments? The debate has been going on for almost 150 years and it is clearly not going to be resolved by reasoned discussion. The democracies where science flourishes are equally dedicated to the idea of religious freedom, provided the particular belief system doesn't lead to practices that involve the abuse of women or children, or are otherwise dangerous and illegal. Surely, if some people want to believe that there is something called 'creation science', that's their business? In any case, my personal perception, from conversations with religious fundamentalists, is that most of them are much more attached to the value they place on traditional moral, ethical and behavioural models than with the creationism. In fact, some of the better educated are clearly embarrassed by the creationist obsession. Is signing on to creationism perhaps a test of faith, a requirement to abandon reason?

The difficulty is primarily in the United States where there is a strong, ground-roots movement in some parts of the country to force so-called creation science onto high school science teachers. That has to be opposed in every place and at every opportunity. 'Creation science' fulfils none of the criteria for science that I describe in chapter 2: it is a belief system and, as such, belongs in the church pulpit. Anyone who holds to these types of beliefs is likely to be a regular churchgoer. There is nothing to stop the fundamentalist churches from inculcating creationist views within their own walls and gardens, so why this drive to invade the schools?

My guess is that this non-science is a Trojan horse for something else: the desire to allow religious observances in public schools, or to gain public tax dollar support for religious schools, or both. The magnificent US Constitution, which was drafted in the spirit of the European Enlightenment, mandates a division between church and state that has been interpreted by the US Supreme Court as requiring a 'wall of separation'. As with many aspects of American law, this is enforced with great rigidity. A country that is based in both a strong Christian tradition and the philosophy of the founding Pilgrim Fathers forbids the singing of Christmas carols, the display of Christmas trees and any form of religious observance in public schools or other state-owned facilities. The denial of what they see as their heritage acts as a point of friction that irritates many rather normal and decent Americans.

The fear in the minds of those Americans who are secular, or who come from non-Christian religious backgrounds, is that the constitutional separation of church and state will be eroded by the powerful political forces currently arraigned against it. The lines in most European societies and in Australia are much less definitively drawn though, as I discussed earlier, these are in many ways societies that can be legitimately described as 'post-Christian'. My own, perhaps somewhat idiosyncratic, view is that the less formal separation between the religious and the political spheres has actually facilitated, rather than prevented, this erosion of belief.

Religion is an enormously important part of the human story. Every school child in the Western demo-cracies should be taught comparative religion and the role that belief systems and practices have played in the evolu-tion of their particular national culture. What happens if religion itself is allowed a voice in public school systems? Australia, which lacks an established church at the level of the nation-state, provides significant amounts of tax dollars for both private religious and secular state schools. In addition, at least when I was a kid, the local priests and pastors came into the Queensland public schools to give an hour's 'religious instruction' each week. Those students who didn't fit into the categories of C of E (Church of England, Anglican), Methodist, Presbyterian, Baptist or Congre-gational had a period off to do something else. It was assumed that Catholics attended their own separate schools. These days, when Australia has a much more diverse population, the public schools need also to allow access to the rabbi, the imam and teachers of other traditions.

It is pretty obvious that teaching 'creation science' alongside molecular genetics, evolutionary biology, geology, chemistry and physics in schools will only cause bright kids to ask, 'Who are these people, and what are they talking about?' Fundamentalists are not stupid, and that, of course, is why they encourage home schooling and set up their own schools and 'universities'. Intelligent young people, no matter what their beliefs, owe it to their development as human beings to expose themselves to the openness and excitement of a good university. Their parents also owe them a level of respect that frees them to make their own judgements. Knowledge enriches. If beliefs are valid, they will survive.

Religion certainly has much less traction in Australia than it does in the United States. Though there are still very committed groups of believers and fundamentalism is gaining some ground, the majority show little deep adherence to religious beliefs—though they embrace ethical and moral values that are broadly Christian, or perhaps more correctly based in Athenian democracy and Aristotle's ideas of more than 2,300 years ago. Many describe themselves as 'Catholic' or 'Orthodox', but often these are as much ethnic and cultural as they are religious identifications. Melbourne, for instance, is the third biggest Greek city in the world. The Greek Orthodox church supplies a connection and a tradition that is life-enhancing for members of this community, but, like most of the long-established Christian denominations, they are losing their young people when it comes to regular church attendance.

The type of Protestant religious culture that was prominent in Western Europe and later dominated colonial Australian life was relatively open, committed to the general good and favourable to the development of science and research. Pragmatism rather than intellectual introspection has tended to characterise the British philosophical tradition. The Anglican Church—Episcopalian in the United States—went through an immensely bruising conflict with science during the nineteenth century, beginning with the famous confrontation between the naturalist Thomas (T. H.) Huxley and Bishop Sam Wilberforce.

The point of contention was Charles Darwin's ideas about natural selection and evolution, published in his 1859 account *On the Origin of Species*. Darwin himself had been studying for the Anglican priesthood when he accepted the invitation to sail as naturalist on the 1831–36 voyage of the *Beagle*. The scientific conclusions that he reached concerning life forms on the isolated Galapagos Islands led him to formulate the theory of evolution, which, as I've discussed earlier, is the basis of modern biology. Bishop Wilberforce was known as 'Soapy Sam' because of his skills as an ecclesiastical debater. Speaking with great eloquence at the British Association for Science meeting at Oxford in June 1860, the good bishop is reputed to have smiled as he concluded his remarks by asking Huxley if it was through his grandfather or his grandmother that he claimed descent from a monkey. Huxley evidently turned to Sir Benjamin Brodie who was sitting next to him and exclaimed, 'The Lord hath delivered him into my hands'—a little irreverent, since Huxley did not subscribe to a theistic view of the world. His response to Wilberforce was:

> I asserted—and I repeat—that a man has no reason to be ashamed of having an ape for his grandfather. If there were an ancestor whom I should feel shame in recalling it would rather be a man—a man of restless and versatile intellect—who, not content with an equivocal success in his own sphere of activity, plunges into scientific questions with which he has no real acquaintance,

only to obscure them by an aimless rhetoric, and distract the attention of his hearers from the real point at issue by eloquent digressions and skilled appeals to religious prejudice.

Huxley is considered to have won the debate.

The consequence of Huxley's advocacy was that over the ensuing years enlightened Anglicans, like their Catholic counterparts, progressively gave up the idea that the biblical account provides anything more than a symbolic explanation for the origin and development of life forms and the natural world. As I discussed earlier, this built on the earlier rise of science that resulted from the Renaissance, the Protestant Reformation, then the seventeenth-century philosophical writings of Francis Bacon and the Enlightenment. Thus, while many English scientists were essentially agnostic, others did not have a particular problem reconciling their science and relatively open forms of religious belief.

Like the United States, Australia has been greatly influenced by the waves of Irish, then Italian, Catholics who settled there. My understanding is that the Catholic hierarchy has no particular problem with evolution or with modern science in general, providing the activity does not cut across the 'sanctity of life' issue that is a central article of faith and forms the basis of the implacable opposition of the church hierarchy to abortion and research that uses embryonic stem cells. If this is a personal moral and ethical concern for someone considering a career in biomedical research, it is important to realise that the great majority of such science does not require any interaction whatsoever with human embryonic material. Apart from that, even those involved in stem cell research who don't have the same moral reservations are working towards strategies, like using post-partum umbilical-cord blood, that will obviate any need to rely on human foetal tissues for future therapeutic applications.

For a long time, the Catholic church held onto the notion that religious doctrine can explain the physical nature of the world. In 1633, Galileo Galilei was forced to recant his belief that the Earth circled the sun, and was not granted a papal pardon until more than 350 years later. Nevertheless, the Catholic Church is justly proud of its intellectual and scientific history. Even in the seventeenth century, the move against Galileo may have had more to do with Church politics than with suppressing his discoveries. Many of the great European universities—Oxford, Cambridge, the Sorbonne, Glasgow, for example—find at least some of their origins in ecclesiastical institutions, and the Catholic Church maintains some substantial US academic institutions to this day—Notre Dame at South Bend, Indiana, St Louis University in Missouri and Loyola University in Chicago. In Australia, there is a network of teaching universities known as the Australian Catholic University.

I was reminded of the scholarly face of the Catholic tradition when I was awarded the 2000 Mendel Medal by Villanova University, a prominent Philadelphia institution run by the Augustinian Order. Gregor Mendel was an Augustinian monk credited with founding the modern science of genetics. His seminal breeding experiments with sweet peas were done at the monastery in Brno, Austria, published somewhat obscurely in 1866, and rediscovered in 1900, with full credit to him, by a trio of European botanists.

The citation for the Mendel Medal reads: 'The Mendel Medal is awarded to outstanding scientists who have done much by their painstaking work to advance the cause of science, and, by their lives and their standing before the world as scientists, have demonstrated that between true science and true religion there is no intrinsic conflict'. Aside from feeling more than a little inadequate as a recipient, I had no quarrel with the high sentiments expressed. The problem is that although it is easy to agree on the universal characteristics of 'true science', as I discuss throughout this book, many of those who are religious have great difficulty reaching any globally inclusive definition of 'true religion'. The degree of flexibility within a particular belief system is clearly a key issue for anyone with deep religious convictions who is contemplating a career in science.

Fundamentalism has been part of the Australian experience for many of the same reasons that it became established in North America. The European nations of the nineteenth century and the early twentieth century adhered to one or other of the established religions, but were often too tolerant for (and less tolerant of) those with a more rigid and puritanical view of the world. Such groups tended to emigrate. Added to that, the twentieth-century reconciliation of schisms in some of the major churches, like the divide between the fundamentalist Free Church of Scotland and the more liberal and powerful Established church (the Presbyterian Kirk), resulted in an excess of pastors, church buildings and a rump denomination (known as the 'Wee Frees') that is now found mainly in the Scottish Highlands. Deconsecrated Presbyterian churches abound in Scotland, serving roles as diverse as tearooms and tourist gift shops. When we lived there more than thirty years ago, one even housed a pack of hunting hounds. The surplus pastors, on the other hand, emigrated.

Fundamentalism is increasing its influence in Australia as, in these very uncertain times, some seek a worldview that is more stable and enclosed. However, fundamentalism does not play out politically in anything like the same way as in the United States. The fundamentalist Family First Party is small and gained a seat in the Australian Senate only because of peculiarities in the preferential-plus-proportional representation voting system. From what I've read, Family First seems genuinely concerned about families, including poor families. They are also focused on helping the homeless, drug rehabilitation, child protection and valuing the elderly. I couldn't find a single word about science or creation science in their published manifesto, and Family First seems to be broadly pro-environment. Recently, it has also been encouraging to see the Evangelicals of the US Southern Baptist convention embracing the idea that they have a responsibility for the natural world as the "Stewards of God's Creation".

There are two likely reasons why a well-organised, extremist fundamentalist view has relatively little political sway in Australia. One is that the majority of Australians disagree with the hard-line 'right-to-life' position of fundamentalist Christians. Although many Australians are uncomfortable about abortion, when it comes to the right to choice, the majority are definitely in favour of this. When more than 4,000 Australians were asked to respond to the statement, 'A woman should have the right to choose whether or not she has an abortion', the answer was in the affirmative for 93 per cent of non-believers, 77 per cent of those with some religious affiliation, 70 per cent of Catholics and 53 per cent of evangelicals (Baptist, Lutheran, Pentacostals). The other factor is Australia's compulsory voting system. Unless they register for exemption on the grounds of religious belief, every adult Australian who is living in the country must turn up at a polling booth on election day and collect their voting paper. They can spoil the ballot paper if they don't wish to register a vote, but they must participate in the process. Compulsory voting was introduced in 1922 when the voter turnout was only 59.2 per cent—which would be regarded as a pretty good figure in a contemporary US election—and has been around the 91 per cent level ever since. This means that it is much more difficult for a well-organised, minority group to gain political traction in the governing house, the House of Representatives, though minority political parties have often held the balance of power in the upper house, the Senate, which is the house of review. Australians should, I believe, protect compulsory voting at all cost. Surely nothing serves democracy better than a high voter turnout?

Other religions also seem able to live in reasonable accord with science, despite popular stereotypes that prevail in Western thinking. Religion is a major force in nations where Islam is the dominant belief system. The events since September 11, 2001, including the invasion of Afghanistan and the ongoing struggle in Iraq, have had the perverse effect of educating those in the West about the depth and power of religious tradition and religious leaders in such communities. Some of the television pictures and stories that many have seen originating from Afghanistan and northwestern Pakistan describe both towns and social structures that seem essentially mediaeval. Guns are everywhere. Unlike the situation in a Western country, the focus of power is still very much the family, with family rivalries assuming the same attributes we recognise from the Montagus and the Capulets of Shakespeare's *Romeo and Juliet*. Of course, this describes only the outer fringes of the Islamic world. A visit to a modern Islamic nation, like Malaysia, quickly leads to the realisation that Malaysians see no conflict in simultaneously embracing religion, science, technology and innovation.

Educated and thinking people in Islamic societies are rightly proud of the open and ecumenical spirit that led to the intellectual, scientific and literary leadership provided by Islam when Christian Europe was in the dark ages. Arabic scholars preserved many of the ancient Greek texts, made useful inventions like the portable astrolabe, and introduced the numerals we all use today. Sometimes it is possible to find a way back to an earlier positive tradition. A hopeful example of this happening is the recent magnificent reconstruction of the Bibliotheca Alexandrina, the Great Library of Alexandria that most believe was destroyed by fire more than 1,500 years ago. The funds to build the Bibliotheca Alexandrina have largely been provided by the governments of Egypt and the neighbouring Arab nations.

Buddhism doesn't create obvious problems for someone who wants to be a scientist. In an opinion piece for the *New York Times* in April 2003, Tenzin Gyatso, the 14th Dalai Lama, writes about his long-term discussions with scientists in fields as diverse as cosmology and neuroscience: 'It may seem odd that a religious leader is involved with science, but Buddhist teachings stress the importance of understanding reality, and so we should pay attention to what scientists have learned about our world through attention and measurement'. He has encouraged scientists to study the physiological basis of the meditation techniques that promote happiness and 'inner balance'. His point is that understanding how meditation works in the scientific sense could help to promote therapeutic approaches that emphasise behavioural change rather than drugs in combating depression and anger. The idea that insights drawn from religion and science can come together to achieve harmony is one that any sane society should be happy to embrace.

The Judaic heritage that has nurtured the early lives of many Nobel laureates, particularly in medicine, is also quite open to the scientific traditions of speculation, experiment and conclusions based in verifiable evidence. The complete list of Jewish laureates includes more than 120 people, though many of these would, I suspect, be essentially secular in outlook. As I understand it, a central idea in Judaism is that, though the words in their holy text (the Torah) stand at the centre of belief, they are not to be taken literally in a changing world, but require constant thought, interpretation and commentary. The result is an intense rabbinic dialogue that depends on questioning and argument, both of which are central to modern science. Claude Cohen-Tannouiidji, the 1993 Physics laureate who grew up in a devout Jewish family in Algiers, summarises this tradition as an emphasis on studying, learning and sharing knowledge with others. Perhaps he could also have added questioning.

This cultural model was greatly reinforced in nineteenth and early twentieth century in continental Europe, where Jews were often excluded from the learned professions like medicine and the law, and the brightest young men became religious scholars. The dam burst following the mass emigration, particularly to the United States and New York, that was driven by events like the Russian pogroms and the rise of Nazi Germany. Australia also benefited, and a good number of the country's leading business people, philanthropists and academics have some Jewish heritage.

What if the question is turned around, though, and we ask how scientists approach religion? In a widely quoted 1998 study, published in *Nature*, Edward Larson of the University of Georgia surveyed the religious beliefs of those who may generally be regarded as the great, living US scientists, the members of the National Academy of Sciences (NAS). Larson concluded that only 5.5 per cent of the biologists, 7.5 per cent of the physicists and astronomers and 14.3 per cent of the mathematicians believed in a personal god. (Interestingly, mathematicians tend to deal with abstractions, and are often not much interested in data.) Among the broader scientific community in the United States, 60.7 per cent are said to be 'doubters'.

Having served on election committees for the national academies of Australia and the United Kingdom (the Royal Society of London) and having been asked as a foreign member for input into the process that selects NAS members, I have no sense that a person's religious beliefs are ever raised or even remotely considered. The inescapable conclusion is that many people who achieve a high level of success in the natural sciences particularly are either agnostic or atheistic to begin with, or they gravitate to one of those positions as their careers progress. There is nothing in the process itself that denies membership to a religious person.

This is not to say that a lack of religious belief is the mark of all successful scientists. Bill Phillips wrote about both his religious upbringing and his commitment to an ecumenical Methodist communion in the biographical account that he provided as a co-recipient of the 1997 Nobel Physics Prize for laser cooling. He had been elected to the NAS earlier that same year. Though he has yet to be awarded a Nobel Prize, another NAS member Francis Collins, the medical scientist who headed the tax-payer-funded component of the human genome project, has no problem speaking publicly about both his strong belief in a personal god and his commitment to the theory of evolution. Coming from a rationalist Protestant background, he was strongly influenced towards a more religious viewpoint by the writings of C. S. Lewis. I fear that reading C. S. Lewis had exactly the opposite effect on me.

Francis has been directly responsible for identifying a number of genetic abnormalities associated with human disease and is a prototype for the concerned research investigator. His viewpoint is: Who are we to criticise a god who has chosen to work his miracles through natural selection and evolution? This is a perspective acceptable to many Christians, though is unlikely to appeal to those who believe that the words of the Bible are God's literal truth.

I know a number of first-class scientists who are, at least occasional, church-going Catholics. One NAS foreign member, the Australian neurophysiologist and 1963 Medicine laureate Jack Eccles was, among many other distinctions, a Papal Knight. He ended his speech at the Nobel banquet with 'May God bless you'. Eccles was a good friend of the philosopher of science Karl Popper. They published a book together (*The Self and its Brain*), which I found to be hard going. Popper's own book, *Conjectures and Refutations: The Growth of Scientific Knowledge*, on the other hand, is much more accessible, but there are no Nobel Prizes for philosophy. In his more speculative writings, Eccles explored an idea he called 'dualist interactionism'. As I understand his position, he believed that the 'self-conscious mind' draws on the physical brain rather than, as most biologists would think, the mind being a product of the brain. Speaking with people who knew him, I gained the impression that Eccles moved away from this idea in later life. He lived to the age of 94, spending his latter years in Switzerland.

After living in both Australia and the American south, and experiencing life on the inside and outside of religion, I can say that good people are sometimes deeply and conventionally religious, and just as often not. Scientists themselves should be careful not to make the mistake of falling into some form of narrow, intolerant, secular fundamentalism. It isn't our job to disparage other people's very personal beliefs, though we are required to present a cogent case for what the evidence says in situations where religious adherence can lead to disastrous results. An obvious example is when religious convictions demand that a child in need cannot receive a life-saving blood transfusion. In such situations, of course, the courts will back the medical professionals against the parents, though this was a hard-won battle.

Scientists do have an obligation to inform and to try to influence public perceptions and policy in areas like disease prevention, vaccines, environmental protection and so forth where there is data indicating that there are real problems that must be addressed. However, though we may sometimes doubt the integrity of religious groups when they attempt to define themselves as the moral and ethical guardians of society, this role doesn't fall particularly on the scientists either. As they mature, people choose for themselves whether to take a more secular or religious view of life. How this equation works for any one individual obviously depends on where they are born and grow up, family influences, experience, education and, perhaps, an inherent spirituality. Sometimes the journey of self-discovery can be very painful, as related by Karen Armstrong in *The Spiral Staircase*. The job of the scientist in this is to present the case for a view of the world oriented to verifiable, evidence-based reality. Any power that we may have rests in the validity and the relevance of the truths that we communicate.

One analysis of the revival of fundamentalism is that the scientific case is so strong that the more dogmatic religious communities fear for the survival of their belief system and in response retreat into a narrow literalism that emphasises the denial of evidence and open higher education. Considering what has happened to institutional religion in Western Europe, there can be no doubt that they have a point.

It is, however, a mistake to regard evangelicals and other religious fundamentalists as representing some sort of monolithic bloc. In the American South particularly, the large evangelical churches, both black and white—for there is a division along these lines—are as much community as religious organisations. Unlike Australia and most of Western Europe, the United States has never accepted the idea that there should be a cradle-to-grave welfare system.

Some of this support role therefore falls to the churches, a trend that is being further fostered under the presidency of George W. Bush. After all, these are tax-exempt organisations, so it is not unreasonable to think that, at the very least, they should take some responsibility for their own communities.

A recent survey of members of these big, mall-like churches, some of which have congregations in excess of 20,000, revealed that less than 30 per cent take their belief beyond regular, or occasional, Sunday attendance. Having lived in this part of the world and knowing a number of such people, I have no particular sense that, even among the literalists who say they believe every word in the Bible, most of them have actually read the book that they claim to base their lives around. Even fewer seem to have any serious interest in theology. Despite that, many derive a great deal of psychological and emotional support from their church involvement. In these difficult times, church membership may be a lot cheaper than psychiatry.

Sometimes belief systems can go in directions that seem both dangerous and irresponsible. When it comes to preventing HIV/AIDS, the mantra that works is the ABC protocol: Abstinence, Be faithful and, failing that, use a Condom. The churches have no problem with 'A' and 'B', but some are directly opposed to 'C'. My friends in the behavioural sciences, who pursue what are termed harm-reduction policies to try to minimise the impact of AIDS in Africa, tell me that the most important thing to do in such situations is to maintain an open dialogue. Perhaps the religious leaders can be persuaded to minimise the extent to which their opposition to 'C' is stated from the pulpit; if that is not possible, maybe they will agree to limit their proselytising to their own community and refrain from promoting a more public stance that will only serve to confuse people generally. The important thing in such a situation is that medical professionals and religious leaders are able to talk to each other in an atmosphere of mutual respect. This proved to be very important in reversing the recent breakdown of both the poliomyelitis and the measles vaccination programs in Nigeria.

Scientists who choose to be involved in the public arena must be prepared to work towards building consensus concerning the issue that they want to promote. This isn't done by being some sort of secular pope or Oliver Cromwell, but requires dialogue, insight and discussion. Certain religious communities will see some issues related to science as important, or dangerous, but others will be welcomed or get an essentially neutral reception. No medical missionary, for example, is going to be upset by the sudden availability of a new and effective AIDS or influenza vaccine, even though the underlying science will be oriented towards defeating the consequences of the rapid natural selection that features so prominently in these infections. If religious groups can be convinced to come in on the right side, they can be powerful allies. We shouldn't automatically assume that even those in the most fundamentalist communities will be firmly in the camp of the evidence-deniers on all issues.

Surely people of good sense who approach the world from a more faith-based religious perspective should be able to agree with those who hold firmly to ideas based in scientific discovery and verifiable reality that the survival of our species, along with the animals and plants in the green and pleasant world that supports our physical and spiritual existence, is of great importance. Whether or not we accept that the responsibility derives from a divine mandate or from the evolutionary need to ensure continuity, it is essential that we all come together to accept that the stewardship of the planet is in the hands and minds of human beings. Though the underlying convictions may be different, it is incumbent on all of us to work towards positive results.

What greater betrayal can there be of God's good grace, or the continuity of our species and all life, than to embrace polarised attitudes of mind and practices that compromise the lives and opportunities of the generations that are to come? If I could ask one thing of my religious friends, it would be that they look hard at political parties and their policies from the viewpoint of global sustainability when they vote, particularly in national elections.

8

Discovering the Future

The morning after the presentation of our Nobel Prize in 1996, Jonathan Mann from CNN hosted a 'Nobel Minds' event. It had been a short night, with few of us being in bed before 3 am. The somewhat spaced-out new laureates sat around in a semicircle and answered questions on challenges for the twenty-first century. The most obvious question was, 'What's next?' I think I talked about solving problems like AIDS and world hunger, but several of the chemistry and physics people turned the conversation to biology. The chemistry laureate Rick Smalley spoke, for instance, about the potential applications of micro-machines and nanotechnology in medicine.

Nanotechnology deals with building minute molecular machines at dimensions smaller than those of a human hair, and micro-machines might be more like a tiny chip and associated 'wiring' that could, for instance, be implanted to help to re-establish the connection between brain and muscle after a nerve is severed. We are all familiar with larger, plumbed-in gadgets like heart pacemakers and the 'bionic ear', so this will, in a sense, be a continuation of an established theme. Nanotechnology is new and at a much more speculative stage. Scientists are working on ideas like building tiny molecular devices that could be injected into the blood to cut away the accumulated cholesterol plaque on arterial walls, or would go directly to disseminated (metastatic) tumour cells and kill them. Think of a scenario similar to the miniaturised submarine and crew in the movie *Fantastic Voyage*, then think millions of times smaller.

Though there are these exciting advances like nano-technology that are just beginning to happen now, scientists are really no better at guessing the future than anyone else. Most specialists can speculate about the long-term consequences of established trends, but novelty and radical change can take everybody by surprise. Humanity has had to deal repeatedly over the past 500 years with revolutionary advances that were both sudden and transforming—just forty years ago, for instance, nobody predicted the Internet, or suggested that many businesses would now be using a globalised, electronically linked workforce. Some of the challenges for the health and longevity of both individuals and the planet are, however, already obvious and ominous.

Like everyone else, when it comes to sooth-saying, anything useful that I have to say is likely to be in the area of my own broad interests: biology and medicine. My fellow 1996 laureates were right: there is potential for immense human benefit from the work being done in biology, particularly genetics and molecular medicine. In fact, when people look back on the science of the twenty-first century my guess is that they will talk of it as the century of biology and the chemistry and physics associated with biological processes. That isn't to say that there weren't enormous advances in these interrelated areas over the twentieth century—there were: from transplant surgery, to antibiotics to heart drugs, to magnetic resonance imaging

(MRI) and so on. However, the biggest changes affecting the human condition probably came from the application of the physical sciences and engineering to transform the realities of transport, labour, communications and, regrettably, weapons.

Looking forward in biology, it is hard to see the excitement diminishing for many years to come. The challenges are enormous because we now have the potential to go from the detailed observation of biological processes to broad landscapes that encompass the chemistry and interdependence of all life. On the one hand, we can progressively see unfolding the nature of precise molecular events that can lead to the development of much better and more specific chemotherapeutic agents (drugs). On the other, it has also become possible to tackle the complexities that are central to the operation of immunity, the brain, human connectivity, and the interaction between all life forms (the biota) and the air, water and earth of our planet.

There is great excitement in all areas relating to genetics, and I don't see that diminishing over the next hundred years. Here is an example as to why: recent research by Jim Downing, Bill Evans, Mary Relling and colleagues at St Jude Children's Research Hospital used genomic approaches to develop a new analysis of acute lymphoblastic leukaemia (ALL). This is a bad cancer of the white blood cells that was a death sentence for more than 90 per cent of the kids who were diagnosed. Progressive advances in chemotherapy and radiation therapy that were pioneered at St Jude brought the survival rate to more than 80 per cent, but there is still a very distressing tail. The mission of St Jude Hospital is 'No child should die before its time', so there is a way to go with this disease.

The scientists and physicians who started St Jude in the 1960s thought ahead. The basement is full of rows and rows of deep freezes containing blood and cancer tissue from every child who has been treated there over the past forty years. The hospital also monitors patients from the time of treatment through to a 15–25-year survival period, so there are very comprehensive clinical records. Our genetics research team went back to these samples and their associated histories.

The messenger RNA (mRNA) that carries the instructions for making the proteins, the building block and regulatory molecules of the cell, is first extracted from thawed tumour tissue, then overlaid on commercially available gene chips to determine which particular DNA segments are being read out in the individual cancers. This information is then correlated with the case histories to give a combined genetic and clinical fate map. The medical application is just beginning, but the long-term consequence will be that gene profiles providing enormous insight for the physicians will now be determined for every new ALL case by a rapid, simplified gene-screen within a day or two of the time that a very sick child arrives at the hospital.

My expectation is that within ten to twenty years, a first step for everyone who is treated in a world-class, comprehensive cancer centre will be to have a blood or biopsy sample taken before the commencement of therapy so that treatment can be individualised to the genetic profile of both the patient and the particular cancer. Though every tumour may differ in specifics, the ALL study suggests that there will be common genetic themes that link different 'families' of cancers that can be treated in different ways. Also, knowing the individual's own genetic make-up will tell the doctors which drugs can both be tolerated and are likely to be effective, and at what level. Researchers throughout the world are currently gathering similar genomic information for all the major tumours of the brain, breast, gastrointestinal tract, reproductive system and so forth. Unlike paediatric ALL, these are high volume tumours that don't require an interval of forty years to accumulate enough samples and clinical histories.

The gene chip referred to above for the ALL study is a direct product of genomics, the science that really got under way with the sequencing of the complete human genome, the major milestone achieved right at the end of the twentieth century. The Affymetrix company, for example, sells two small chips (looking much like those in the back of a digital camera) that express the 30,000-plus genes that provide the necessary code to make a human being. We have, of course, no idea how to do that other than by the traditional way of having a sperm fertilise an ovum. Most sane human beings would not want to see that particular situation change. Think, all the same, how great it would be for transplantation if we could take some stem cells from an individual's own bone marrow and use our understanding of the particular sequence of molecular interactions that determine organ development to make a new kidney or pancreas in the laboratory. We are decades, perhaps centuries, away from being able to do anything remotely like this.

What will undoubtedly happen over the next twenty years or so is that the application of molecular genetic and genomic approaches will lead to the development of novel anti-cancer drugs that are less toxic and have fewer side effects. The first so-called designer drugs have already been generated by applying strategies using structural chemistry and molecular modelling. What this means is that the nature of an interaction between, say, two proteins that come together in the cell cytoplasm to provide an on/off (divide/stay quiet) switch within the cell is sorted out by the molecular biologists, who work in typical 'wet' laboratories with incubators, microscopes centrifuges and the like. The proteins are then crystallised from solution under specific chemical conditions and, once a suitable co-crystal of our two 'switch' proteins is formed, the complex is passed on to the structural biologists. Their job is to generate 'pictures' of the molecular 'surfaces' that tell us exactly how the two proteins interact.

The structural guys traditionally used X-ray crystallography, the technology that was first developed by William and Lawrence Bragg (Chemistry, 1915), then refined to illuminate biological systems by a whole range of Nobel Prize winners, beginning with Max Perutz, John Kendrew and Maurice Wilkins in 1962 and Dorothy Crowfoot Hodgkin in 1964. The 'new' structural biologists, like Rod MacKinnon, who won the Nobel for Chemistry in 2003, use the smaller linear accelerators, the synchrotons (which I referred to earlier in the book), for the same purpose, a change that enables the analysis of more complex interactions and infinitely speeds up progress of this area of science. Many structures are now solved with a rapidity that would have looked impossible to Rosalind Franklin; she took the key X-rays of DNA that allowed Watson and Crick to sort out the double helix structure, but died of cancer soon after that, in 1958.

Knowing the topography of the molecular interface then allows various types of chemists to design and test small molecules (drugs) that might fit into the binding site of one or other molecule to block the interaction. This complex and difficult chemistry is very much helped by modern computer simulation techniques. Any young person who is obsessed with computer design might think of studying chemistry. A note for the squeamish: not all medical research involves dealing with blood and guts! Apart from opening a new world of insight into patterns of extraordinary beauty and elegance, this is a very exciting, complex and satisfying area of scientific research that has enormous potential for enhancing human well-being. The computer modelling side of this activity is what we call 'dry lab' science. Another big growth area that uses heavy-duty computing in the 'dry lab' medical world is in the new mathematical and statistical science of informatics, the discipline concerned with sorting out the huge volume of complex, novel information that is being generated from the 'gene chip' type approaches of genomics.

The first anti-tumour agent to become available as a result of the rational drug design approach that I've described above is Gleevec, an inhibitor that binds to a particular molecular target (a tyrosine kinase enzyme) and blocks the uncontrolled cell proliferation that is characteristic of cancer. Until now, cancer chemotherapy has essentially depended on the use of poisons to kill the dividing cells. The result is hair loss and destruction of the immune system, which makes patients very susceptible to infection. A much more specific therapeutic like Gleevec lacks these side effects.

The problem that is emerging with Gleevec is the same as with any cancer therapy: the development of resistance. Before discussing that, it is necessary to say a little about the nature of cancer. The following discussion is very simplistic, as there are many types of cancers that can develop for different reasons, but it should be sufficient to provide some understanding of how tumours escape from drug control.

With the exception of a few cancers that are caused by viruses, tumours develop as a consequence of genetic mutation(s). A mutation is simply an error within a gene that occurs at the time of cell division and, providing it isn't lethal, will then be passed on to 'daughter cells' with, in most cases, no undue consequences. This process, called 'background, somatic mutation', is occurring all the time in every one of us, but there is generally no reason to select for the particular clone of cells that carries the altered gene. Cancer develops when the mutation disrupts the signals responsible for the normal process of growth control, the progeny cells continue to divide and, depending on how fast and where this growth proceeds, the characteristic lump is eventually detected. Often one or more sequential mutations are required to trigger the process. This is where ultraviolet light (skin cancer) or the coal tars in cigarette smoke (lung cancer) play a role as 'co-carcinogens' by inducing DNA damage and mutations.

In genetic terms, tumours are both selfish and dumb. The tumour genome is 'selfish' in the sense that it takes no account of the needs of the tissues, the organ or the sensate being that surround it. Due to genetic changes, it has broken free of the normal 'social controls' of the body, like contact inhibition. It is, in a sense, a cellular entrepreneur gone mad. The tumour genome is 'dumb' because it cannot 'know' that by killing the individual that provides the nutrients to support its own survival, it is also committing suicide.

Because the tumour genome is completely selfish, it will mutate and try to defeat any type of inhibitory mechanism. Gleevec, or any therapeutic agent, supplies an additional evolutionary 'pressure', favouring the selection of mutant clones that have managed to engineer a further escape from the drug. The solution in conventional cancer treatment is to use multi-drug chemotherapy, where each chemical targets a different molecular mechanism within the cell. This is the high-wall, prison-cell, ball-and-chain approach to containment. It is much less likely that the cancer will be able to develop mutational changes that defeat two or more different control pathways. This diversity of possible treatments will also need to emerge for the 'designer' drugs, like Gleevec, if we are to develop non-toxic, specific cancer therapies that are effective in the long term. Given the rate of progress and the science strategies that are already available, we can expect to see significant advances in this field of rational drug design over the coming decades.

Many infectious agents will also be on the 'designer' drug hit-list for this new century. Such compounds are already being used to counter influenza and HIV/AIDS, though there is always the danger that mutant viruses will escape and no longer bind the particular chemical inhibitor. Apart from the selective pressure applied by the drug, mutant HIV and influenza viruses are constantly emerging as a consequence of the need to defeat immune control. When it comes to viruses, we are dealing with biological entities that cannot be either totally selfish or dumb.

Viruses, unlike tumours, are not 'dead end' in nature. In order to grow and survive, they must be able to exit the body and transmit to other hosts. This requires the operation of many different molecular mechanisms, any of which might be damaged by a mutation that would simply allow the virus to escape drug or immune control within a particular individual. The 'designer' drugs Relenza and Tamiflu, for instance, specifically inhibit influenza virus growth by binding to a viral surface protein, the neuraminidase, which normally functions to allow escape from the cell it infects. Both these chemical inhibitors have now been in use for some years, and there are no signs that highly infectious, resistant mutants are emerging. If the lethal H5N1 avian 'bird flu' adapts to transfer from human to human, Tamiflu should provide an immediate first line of defence (see chapter 4).

It would be possible to develop a whole range of such anti-viral agents that target, for example, the parainfluenza viruses that cause croup in young children. There are, however, three problems. The first is that the rapid course of any respiratory infection allows only a small 'window' between the onset of symptoms and the time in which the therapy will be useful. This is worthwhile with a potentially lethal infection like influenza, as early treatment can mean the difference between survival and death. Such drugs can also be taken beforehand as a preventive measure in the face of a global outbreak to protect, say, the very young and the elderly who are particularly at risk. The second difficulty is in being sure that, as 'designer' drugs are very specific, the right infection is being treated. There are hundreds of viruses that cause debilitating and annoying respiratory infections, so any expansion of this type of therapy requires the development of rapid diagnostic tests. The third—and perhaps the biggest problem—is in the economics relating to the cost of the drug and its development and testing. The same constraint applies when we think, for example, about making a 'designer' anti-cancer drug to treat some rare paediatric tumour.

We already see the effect of economics in the shape of the current AIDS pandemic. The use of expensive triple drug therapy in the advanced world means that contracting this infection is no longer an immediate death sentence— though there is no cause for complacency, as it is still a shocking disease, and the daily ingestion of anti-viral chemicals can have major side effects that shorten life. The lack of financial resources to pay for the drugs and to provide the necessary infrastructure means, however, that the mortality rates in the developing countries are horrendous. Whole societies are losing many of their teachers, farmers and public officials. Planning for the future is impossible in situations where the survival of both parents and children is uncertain. The distribution of cheaper generic drugs is now more widespread, but they are still unavailable to many, and the AIDS pandemic remains an enormous challenge for humanitarians. On a brighter note, the experience in Thailand, Senegal and Uganda tells us that it is possible to promote behavioural change and reduce the toll.

The AIDS challenge for medical scientists is to develop affordable, effective preventive measures. All types of approaches are being tried, from the topical application of lemon juice and more sophisticated formulations that can be used by women to prevent transmission, to the development of very complex vaccines. As I discussed earlier in the book, the results with experimental vaccines have so far been extremely depressing, and it is likely that some major, new conceptual breakthrough will have to occur before we see even the possibility of a solution. I am personally involved at one level or another in three different programs, two in Australia and one in the United States, and there is nothing that looks even remotely like a certainty.

One problem is that, unlike the influenza virus that is completely eliminated within one to two weeks of exposure, HIV both persists and mutates to avoid immune control. Another is that there is no mouse model of AIDS that we can use quickly and cheaply to try new immunisation strategies, while the HIV-like viruses that infect non-human primates seem not to provide a perfect mimic of the human disease. Even so, this is the best we have and it is unlikely that a candidate vaccine showing little protective effect in preliminary monkey studies will be of much value for people. The result is that the only true test of a novel AIDS vaccine is to do human trials in endemic areas where the disease is currently spreading. Such studies are politically and strategically complex, and very expensive.

This is where I will spend the remainder of my scientific career, trying to develop a better understanding of what is going on during the course of immune responses to both influenza and HIV infections. I referred earlier to a recent WHO estimate concerning a potential pandemic: it suggests that even with modern drugs and improved vaccine technologies, there could be in the order of 72 million deaths if an influenza virus as severe as the current H5N1 infection in birds breaks across into the human population and causes a global pandemic. At the 2002 mortality rates, which are increasing, that would equate to twenty-four years of AIDS deaths. These are sufficiently important problems to keep me going, as I said before, as long as colleagues think I am still contributing to innovative science.

Another infectious disease problem that is of immediate concern is the rapid development of antibiotic resistance, leading to the emergence of conditions like refractory tuberculosis, 'golden staph' and necrotising fasciilitis, the latter caused by 'flesh-eating' variants of the ubiquitous type A streptococci. Here there is cause for considerable optimism. All the antibiotics in common use are, in fact, natural defence molecules used by a few species of bacteria and fungi. There are millions of different micro-organisms that could potentially provide novel products that operate in quite different ways. Apart from that, a combination of more traditional 'reductionist science' and genomics approaches is allowing us to identify a whole spectrum of new defence mechanisms used by species as diverse as termites and trees. Using modern techniques, the genes coding such a molecule can be cloned, then made to express the protein of interest in bacteria so that it can be screened for possible effects on the various bugs that plague us and our domestic plants and animals.

Craig Venter, the scientific visionary who contributed enormously to solving the human genome in record time, is currently sailing around the world on a yacht sampling all the life forms in the top two metres or so of the sea at various locations. His team negotiates questions of 'ownership' with the various governments that lay claim to the different sampling locations, then uses genomic approaches to probe the total genetic spectrum for the different plankton and so forth that are retained on filters of diminishing pore sizes. Even at this early stage, they have identified what could be the basis of novel photosynthetic mechanisms, raising the possibility that there may be revolutionary, non-polluting ways of solving the sunlight to energy equation. Perhaps a team of innovative molecular biologists could use one of these new plankton genes to make bacteria that convert the energy of sunlight to hydrogen in some novel type of fuel cell. On a more mundane level, what the Venter team is also doing is providing extraordinarily valuable 'biota' baselines against which the effects of global warming in the oceans can be sequentially measured over this coming century.

Not all advances in medicine and so forth will necessarily depend on the use of modern molecular genetics and genomics. China, in particular, has gone back to look very closely at the active principles in their traditional natural medicines. A recent development is the very effective anti-malarial artemisinin, which is currently in short supply because it has to be made from plants growing on Chinese and Vietnamese farms. As related in *Science* (7 January, 2005), the Bill and Melinda Gates Foundation recently committed $40 million to developing genetically modified bacteria that will churn out a precursor of the drug. In addition, knowing the nature of artemisinin has allowed the chemists to make variants that work even better. A former colleague from Canberra days, Graham Johnston at the University of Sydney, is also in the business of identifying the key ingredients in natural herbal medicines. As he pointed out to me, these 'natural medicines' tend to be fairly weak drugs. If this were not so, the variations in growing conditions, life cycle and so forth could lead to differences in the concentrations of the active chemicals in the plant that might result in overdose and toxicity.

The challenge with these drugs of plant origin can be that, even when the elements are identified, the chemistry required to make synthetic analogues may still be too difficult. Again, this situation is likely to improve through the years to come, and is a challenge for aspiring young chemists.

I discussed the future opportunities in my own field of immunology, including the need to develop a range of vaccines that protect against persistent infections like HIV/AIDS, in chapter 4. Similarly, the controversy surrounding plant genetic engineering and the global warming debate was addressed in chapter 2. What all these matters have in common is that they are complex and multifactorial. A big concern for the environmentalists, for instance, is that genetically modified organisms (GMOs) will escape and either displace the natural flora, or transfer disease-resistance genes to weeds. It's hard to see how either effect could be worse than what has already happened with introduced plant species and, as Tim Low points out in his book *Feral Future: The Untold Story of Australia's Exotic Invaders*, there are plant 'time bombs' growing in any municipal botanic gardens that could potentially break out into the wild and cause major problems. This has evidently happened before and will happen again. How might global warming play into such a 'plant/escape' scenario? Growing insect-resistant GM plants should also mean that we now spray less insecticide around the place. More insects could mean more insectivorous birds. Birds carry some mosquito-transmitted viruses, like the Murray valley encephalitis virus, that can cause severe symptoms in humans. The combination of global warming, insect-resistant GMOs, more insects (including mosquitoes) and more birds could result in more human disease. This is just one example of a complex, interactive system.

Dealing with complexity may be the major challenge for science in the twenty-first century. Understanding and manipulating the complexities of infection and immunity, or of cancer immunity, provides a continuing, major research focus for those interested in vaccines and therapy. Genomics and the array techniques of 'discovery science' have opened the way to scan the totality of the complex genetic 'read-out' in a cancer cell or an activated T lympho-cyte. Through the latter part of the twentieth century we had great success in the type of reductionist science that looks at the different parts of various molecular pathways. This will continue, with the discovery of novel genes and molecular interactions forming the base of many new scientific careers. The big questions may be concerned, however, with putting the whole molecular machine together to explain function at the level of the cell, the organ and the organism. As yet, though the triumphs of medical science have been extraordinary, we haven't even answered such simple questions as what makes a liver grow to the shape and size of a liver?

The great complex and mysterious system that defines all of us is, of course, the human brain. Here we can expect enormous advances in understanding over the next hun-dred years. New insights will undoubtedly come from molecular biology and genomics, where individual genes and/or predictable profiles of multi-gene read-out will be associated with both physiological abnormalities and psychosomatic disorders, or mental illness. Deciding what to do with such information may not always be straightforward. Applying new targeted therapies that alleviate debilitating conditions like schizophrenia or epilepsy might seem to raise no obvious problems, but how does society handle a situation where an individual's DNA pattern is associated with, for example, a tendency to extreme violence and criminality?

The question has already been raised by the genetic correlation between monoamine oxidase A promotor (MAO-A) polymorphism (genetic variability) and a tendency to violent, antisocial behaviour for people with a particular MAO-A genotype. We can't lock people up on the basis of a tendency. If possible candidates (or their parents) who have committed no crime refuse to be tested, how is this to be handled both legally and ethically? Perhaps some types of severe crimes could be totally prevented by identifying potential offenders as children then providing appropriate counselling or drug therapy; but does society have a right to insist on such intervention in those who constitute a clear risk? On the other hand, if such a genetic predisposition is first identified after the event, is it legitimate to use severe punishments against people who could be said to be trapped by their underlying biology?

Even the schizophrenia and the epilepsy examples are not as simple as they may seem. Though individuals and their families would clearly benefit, effective treatment may also eliminate certain elements of the cultural dynamics that have shaped human society. Most would agree that we can do without either of these debilitating conditions, just as we have no reason to miss diseases like smallpox. However, though treatment with the appropriate psychoactive drugs might indeed have saved Vincent Van Gogh's ear, would this have been at the expense of his artistic vision and creativity? There is little doubt, I think, that most schizophrenia sufferers would gladly accept any such loss if the medication allowed them to function normally in society.

Temporal lobe epilepsy is frequently associated with episodes of profound spiritual experience. Neurologists have argued that the account of St Paul's revelation on the road to Damascus is, in fact, a classical description of such an epileptic seizure. How different would the history of Western civilisation be if we could time-travel back to the beginnings of the Christian era and give St Paul the twenty-first-century epilepsy therapy that may be just around the corner? What do we lose if a more sophisticated understanding of brain function leads to the implantation of electronic chips, or to an extensive pharmacopeia, that modulates the more distressing and debilitating extremes of the human experience? Again, I believe the majority of those who are subject to epileptic episodes would opt to give up the possibility of a spiritual revelation for well-being and the right to hold a driving licence.

Many, if not all, forms of drug addiction are likely to have a basis in how a particular individual's nerve cells respond to the various neurotransmitters. These are the chemical signals that pass from one nerve cell to another to stimulate the electrical activity that allows the brain to function. A key neurotransmitter is dopamine. Parkinson's disease, which is classically characterised by tremor, a stooping gait and slowness to initiate and maintain movement, is due to the loss of many of the dopamine-producing nerve cells in one particular region of the brain, called the substantia nigra. These patients are treated with a drug that substitutes for dopamine, L-dopa. On the other hand, people suffering from schizophrenia may have too much dopamine and benefit from being given dopamine antagonists that block neurotransmission and thus over-stimulation. Cocaine can inhibit dopamine removal, leaving more around to give continued stimulation. The mono-amine oxidase inhibitor that we met earlier, when talking about the genetics of violent behaviour, normally functions to break down dopamine.

It would not be surprising if the application of modern genomic screening, the gene chips that we discussed in relation to cancer, reveals genetic profiles associated with neurotransmitter levels, sensitivity, and so forth that determine the susceptibility of any given individual to drug or alcohol addiction. Many of these conditions are likely to reflect the interaction of multiple genetic effects or, as they're known in the genetics trade, complex traits. Knowing that someone is born with a genetic predisposition should lead to more focused education and prevention programs, while understanding the nature of the molecular targets is likely to result in the development of better therapeutic agents.

The nature of drug delivery systems should also improve. Nowadays, if we take a tranquiliser or a sleeping pill, the active chemical is distributed through the blood and binds to the appropriate receptors wherever they may be. Perhaps the application of nanotechnology approaches will lead to the development of drugs that are molecular machines destined to go only to where they are needed. It's the difference between using police to find an offender in a big crowd at a football game, or sending a couple of detectives to a house the guy was seen entering half an hour earlier. The process is both more economical and less likely to result in unhappy side effects.

The other big challenge when it comes to brain function is posed by degenerative neurological conditions like the pre-senile dementias, Alzheimers disease and so forth that are an increasingly serious plague in the elderly. The dementia epidemic reflects that human beings—in the developed countries, at least—are living longer than they did even thirty years ago. A lot of this is a direct result of improved preventive treatments for cardiovascular disease. What happens in a condition like Alzheimers is that wrongly folded proteins accumulate in, or around, the irreplaceable nerve cells and eventually poison them: think of the tarry gunk from an oil spill at sea that is washed ashore and chokes the plant and animal life.

One approach is to immunise with one of the offending gunk proteins, called amyloid, so that it can then be removed by the resultant immune response (discussed in chapter 4). However, this has so far proven to be too dangerous. It worked well in mice that were genetically modified to express human amyloid protein in the brain but, for obvious reasons, the initial human trial focused on people with advanced disease. At that stage, there is so much amyloid around that the immune cells that invade from the blood cause severe symptoms as they try to get rid of it. The best solution is likely to be the development of a small molecule (a drug) that will block the folding process. Even if this only served to delay the onset of symptoms, the benefits not only in terms of alleviating human suffering but also in health economics would be enormous.

One certain development through the twenty-first century is that the different scientific disciplines will work more closely together. This is hardly a new trend, but what is new is that major institutions are taking active steps to facilitate interactions between those who come from different science cultures. The history of Nobel Prizes shows how significant this interaction can be: the first Nobel Prize in Physics was awarded to Wilhelm Roentgen for discovering X-rays, and the invention of X-ray crystallography by William and Lawrence Bragg led to structural biology, which has been recognised by a number of Nobel Prizes and continues to contribute massively to biology and medicine. The 2003 Medicine Prize went to the chemist Paul Lauterbur and the physicist Peter Mansfield for the development of magnetic resonance imaging (MRI). Nobel Prizes for Chemistry are frequently given for discoveries and technological developments that have their major application in biology, the two awards to Fred Sanger for protein and DNA sequencing being a case in point.

Breakthroughs often result from bringing fresh minds trained in different fields together for some common purpose. Rolf Zinkernagel had taken a course that emphasised current thinking in immunology, but I had worked previously in virology and pathology and we were both ingénus at the time we did the key experiments and thought 'outside the box' about immune recognition. Established theoretical physicists like Erwin Schrödinger (Physics, 1933) and Max Delbruck (Medicine, 1969) made substantial contributions when they switched their interests to biology.

Max Delbruck worked at Caltech after he fled the Nazis, but in 1945 he also started a summer course on the genetics of bacteriophages (viruses that infect bacteria) at the Cold Spring Harbor Laboratory (CSHL) on Long Island Sound, New York. The bacteriophages were the initial research tools that led to the current era of biotechnology and molecular medicine. The summer courses at CSHL continue and—still an intellectual powerhouse—it is legitimately regarded as the ancestral home of molecular biology. A picture on a wall at CSHL shows a very young Jim Watson, of Watson and Crick fame, working there in a summer job as a waiter, while he was a course participant. Jim later served as CSHL director for more than twenty-five years and was also the first director of the US federal human genome project. He continues in a senior role as CSHL President, and was succeeded as director by the JCSMR-trained Australian virologist, Bruce Stillman. Max Delbruck, Al Hershey and Salvador Luria shared the 1969 Nobel Prize for Medicine, for their 'discoveries concerning the replication mechanisms and the genetic structure of viruses'. Al Hershey worked at CSHL, and Salvador Luria was Jim Watson's PhD supervisor at the University of Indiana.

There will be similar science stories through the twenty-first century, though only a science fiction writer could guess at the fields of interest, who the characters might be and how the stories may unfold. Apart from improving the human condition, science has a job to do in protecting humanity and the world we live in. What could be the ultimate threat? We have to hope that no person or group is crazy enough to start a nuclear war. One theory to explain the extinction of the dinosaurs is that there may have been a massive asteroid hit, creating such a storm of atmospheric debris that the life-giving rays of the sun were blocked out. No doubt the probability is low, but what could, or would, scientists do to prevent a recurrence? Perhaps we couldn't stop the hit, but might it be possible to work out how to be independent of solar energy until the dust cleared? We must continue to stretch both our imaginations and our understanding to the utmost. If we want our species to survive in the long term, human beings cannot afford to stop reaching for the stars.

9
How to Win a Nobel Prize

So you want to win a Nobel Prize: to become famous, powerful and maybe even very wealthy? If that's your ambition I can't help you. There is no instruction manual or course that can guide you to a Nobel Prize and, numerically speaking, most of us have more chance of winning an Olympic gold medal. There's also another difference: an Olympic medallist might go on to win a Nobel, but can you imagine Albert Einstein or Bertrand Russell competing in the decathlon? I was brutally reminded of this when I had to present a large cheque to Michael Chang for winning the St Jude Tennis Classic in Memphis. We were both winners in one sense or another but, though Michael might conceivably change his life at some stage to become a great scientist or writer, there is no way that I could ever beat even an 85-year-old Chang or Sampras on the court.

Now that I've had your attention and you have read this far, I hope you will recognise something of what it takes to make an outstanding research scientist. It involves a personal recognition that humanity advances by insight, discovery and a capacity for serious effort and commitment. So, following what I've written here won't guarantee a trip to Stockholm or Oslo, but, with a little luck, it could lead to something worthwhile.

Try to solve major problems and make really big discoveries

The individual who is well educated, works enormously hard and has inherited extraordinary ability and intellectual capacity might just conceivably be able to identify a major problem at the Nobel level of achievement, then move ahead to solve it. From my experience such people are pretty rare, and may well be either alien life forms or the next stage in human evolution. Discovery is different. Nobody can decide to discover something, but there are ways of making a discovery more likely. Focus on generating new information and insights and look for unexpected outcomes and results. Accept nothing at face value and get in the habit of thinking unconventionally. Work hard, work smart and, with a bit of luck, serendipity will play its part.

Be realistic and play to your strengths

A trained veterinary surgeon like me knows, like all punters, that there are horses for courses. Everyone has to find out what sort of horse they are. Anyone with a brain that does best at ploughing long, straight furrows should give up on the idea of being an intellectual polo pony or steeplechaser. Perhaps a molecular biologist or organic chemist can also be a poet, but it's likely that most will do a lot better at one than the other. Science at its best is for people who love to ask questions and are delighted by discoveries that overturn established ideas and prejudices. If they have to choose between authority and evidence, basic scientists will always go with the evidence. Most scientists are notoriously contemptuous of authoritarian politics, for example. Any love affair between science and politics is always fraught with potential conflict, though the passion and betrayal that characterises tempestuous affairs often makes the best theatre, or press—as is usually the case.

Acquire the basic skills, and work with the right people

The elements that make an exceptional humanitarian or writer can be as varied as the individuals themselves—apart from the obvious ones, like a keen intellect and a serious sense of commitment. On the other hand, scientists have an absolute need for in-depth, specialist training at the university undergraduate level and beyond. Though it's not essential, it helps to be born into an intellectual and supportive family, grow up in the United States, Europe, Japan, Canada or Australia, attend an academic school and a great university, and train with a top person. Some aspirants try very hard to work with a Nobel laureate, as they have the right to nominate people for Nobel Prizes—and only those who are nominated are considered. Senior scientists like to think that they create enduring 'schools', so it can help to be part of such a lineage. None of these factors, however, will guarantee a Nobel. Sometimes the idiosyncratic outsider will rise to the top over those within the big tent of mutual reinforcement, where it can be too warm and too comfortable. Thank goodness for that. Otherwise science would be a stuffy and obsequious business.

Learn to write clearly and concisely

Many people who are very good at science are great doers, but uninspired writers. It isn't necessary to be a Shakespeare or a Michael Ondaatje, but anyone who wants to be recognised as a top scientist must be able to write clear, concise English. English is the language of science and many countries, among them Malaysia and Singapore, that are building their science profiles teach in English at school and university level. Science is about telling good, readable, memorable stories.

Work in an appropriate field

Nobel Prizes recognise some, but by no means every, aspect of what we may think of as the high culture of humanity. There are no Nobel Prizes for the visual arts, for music or for dance, so these might be fields you would want to avoid. The performing arts have been recognised only twice, to my knowledge, by the 1997 Literature award to the Italian playwright Dario Fo and the 1953 Literature Prize to the British leader Winston Churchill, who was, of course, a noted orator as well as a writer—although the public advocacy required of many Peace laureates as they seek to promote their particular interests might be considered partly under the head of 'performance'. For scientists, speaking about their work to both specialist and broader groups is an essential component of receiving credit and building a reputation. You and your work will remain anonymous if you just stay home.

You will also need to exercise caution in the area of science you work in. While some areas of research are not specifically identified as targets for Nobel Prizes, they may slip under the wire in another category. Though pure mathematics may have been excluded, theory based in mathematics is clearly central to physics and economics, as exemplified in the Economics awards to John Nash in 1994 and James Mirlees in 1996. Geology, for instance, is not specifically identified, but it is possible that someone who trained primarily as a geologist might be honoured for contributions to physics or chemistry.

There is no agriculture prize, but plant scientists who have been recognised include the wheat breeder Norman Borlaug referred to earlier, who won the Peace Prize in 1970 for the part he played in what has become known as the green revolution; the agricultural scientist and research administrator John Boyd Orr was awarded the Peace Prize in 1949 for establishing the FAO; and the plant geneticist Barbara McClintock the Prize for Medicine in 1983 for jumping genes in corn. Still, anyone who is set on the idea of a Nobel Prize would probably leave the plants to someone else. Otherwise, there is a world food prize.

Find and cultivate your true passion

Despite everything I've just said in the above, one of the best things that can happen in life is to discover a line of enquiry that really grabs your interest. Someone who has found a passion that doesn't fit the Nobel, or any other mould that conventionally leads to the 'glittering prizes', should forget the award and go for the satisfaction and the excitement of what they love doing, whether it's philosophy or building surf boards. That's where the prizes that really matter are to be found. According to the poet Ezra Pound, who would certainly have been ruled out of consideration for a Nobel Prize because of his fascination with Mussolini's fascism, 'What thou lovest well remains, the rest is dross'. Preoccupation with dross and irrelevance is a sure-fire way to avoid the Nobel Prize. Most of those who do win have not only achieved over a long period, but are likely to have given their full attention, energy and enthusiasm to what they do. On the other hand, a passion for working out mechanisms for usefully recycling the dross of packaging, junked cars and so forth that we produce in our daily lives could lead to one of the numerous environment prizes that can be identified by searching the web.

Focus and don't be a dilettante

Most scientists and economists are identified with a particular sub-field of investigation for much of their lives. Sometimes the best scientists will take on new challenges, but the majority—and particularly the prize winners—tend to remain within the same broad field, like cancer biology, neurobiology or immunology, where they are well known and regarded. Bright people who hop around from one topic to another often achieve very little.

The exception may be the Nobel Peace Prizes. Concentration of effort may be required only in the relative short term for the humanitarian who wins a Peace Prize—the successful resolution of a major confrontation, for example, can depend on an individual's political power or stature as a negotiator. These credentials are likely to have been achieved in a completely different context, such as being US Secretary of State: Cordell Hull, Peace Prize, 1945; Henry Kissinger, Peace Prize, 1973. Other causes, like the elimination of landmines, gain traction only because someone like the 1997 Peace laureate, Jody Williams, is totally dedicated to finding a solution. The novelist or the poet may contribute in a variety of forms, though their style and approach may be consistent. Writing itself is the most intense of human activities. Again, it's all about intensity and commitment.

Be selective about where you work

As a scientist, your chances of achieving anything can be greatly diminished by working in an institution that is under-resourced financially, does not value creativity or demoralises even the bright people that it manages to recruit. The places that nurture winners don't all look the same, and can vary from small private institutions like CalTech, to massive Ivy League conglomerates like Harvard University, to state institutions like Southwestern Medical School in Dallas. Every one is different, so find an environment that suits your personality and work habits. Being in the regular company of colleagues who are stimulating to talk to and living in a culture that values creativity and insight contribute mightily to a satisfying life, even if the big prize doesn't come your way.

Value evidence and learn to see what's in front of your nose

In the end, science is about data and being able to 'read' the real meaning of what you find. Keep an open mind, be prepared to think laterally, and be instructed by nature and observation. Great literature and visionary science share a characteristic: the reader recognises that there is a sense of truth. The resolution of conflict or the navigation of an impasse that has been achieved by a Nobel Peace laureate also reflects a capacity to recognise, then act on the underlying reality. Such people bring to the task not only a substantial intellect, but also a clear perception of what can be achieved.

Think outside the box

Following the obvious path is not likely to lead to a novel question, interpretation or solution. If the way is both straight and narrow, the odds are that somebody else already will have gone down that road. The mind works in strange ways, and it can help to short-circuit, or bypass, normal thought processes. Edward de Bono formalised one such technique when he identified the 'lateral thinking' approach. When struggling with a scientific problem, it often helps to 'draw' the possibilities, either in your mind or on a piece of paper. Human beings think in both words and pictures. Illuminating ideas come at odd times, in the shower, for instance, or on the top of a mountain. Ilya Mechnikov, who won the Medicine Prize in 1908, famously discovered phagocytosis when, bored and at the beach, he poked small thorns into starfish larvae and watched the inflammatory cells congregate at the site of injury. Get rid of the clutter, and let the mind roam.

Physical activity, even if it's only walking, can also work to free the thought-processes. New ideas often seem to pop up when the mind is idling or half-concentrating on some more mechanistic activity, like gardening or building a bicycle shed. The 1993 Chemistry Prize winner, Kary Mullis, describes in his autobiography *Dancing Naked in the Mind Field* how the idea of the polymerase chain reaction came suddenly when he was tired and driving alone at night. This won't happen to you if your brain isn't grinding away at the problem in the background, which only happens if you are very attached to the particular question —obsessed even. Ask anyone who is married to a serious scientist, and you're likely to get into a discussion on the nature of obsession.

Talk about the problem

Don't be a lone wolf. Two heads are often better than one. The most obvious way to develop novel insights is to talk with others, particularly those who come at the issue from different backgrounds. Jim Watson's little book *The Double Helix* gives a strong sense of the intense interaction at Cambridge's Cavendish laboratory between the biologist (Watson) and the physicist Francis Crick (Medicine, 1962) as they tried to build their DNA model. The essential information that they had been using the wrong tautomeric form of the DNA bases came from Watson's chance conversation with a visiting American, the crystallographer Jerry Donohue. Without knowing this they were stuck. Though Rosalind Franklin took the key X-ray pictures that provided the solution for Watson and Crick, she and her King's College, London, colleague Maurice Wilkins failed to develop a rapport. Rosalind remained very isolated and she did not solve the DNA problem.

There is a caveat. If you make a major discovery that could easily be repeated by others, it's best to keep quiet about it until the initial research report is either published or 'in press' in a top journal. Anyone who works in basic science at a high level is likely to have had the experience of the odd elliptical discussion with a colleague who is brimming with excitement about a new discovery but just can't afford to talk about it pre-publication, the I-could-tell-you-but-I-would-have-to-kill-you scenario. This is one time where you need to think like an MI5 operative.

Australia is different now, but Rolf Zinkernagel and I benefited back in the 1970s from the isolation of this 'distant shore'. Talking openly in the local immunology discussion group, the weekly 'Bible class' run by the department head, Gordon Ada, certainly helped to clarify our thinking. That would be much riskier today: Australia is definitely in the scientific loop, and national and global communication means people can hear about a discovery almost the instant it is made. An inadvertent comment in someone's e-mail could provide the necessary clue for a competitor. Once our first papers (which are in the appendixes) were accepted for publication, however, we both did tours of the northern hemisphere and gave an enormous number of seminars to publicise our findings. I was exhausted by the time we met up for a final week at the 2nd International Congress of Immunology in Brighton, but Rolf was still jumping up and inserting the story of our discovery at every appropriate (and inappropriate) opportunity. Someone dubbed him 'HyperZink'. The sooner a major new scientific finding is out in the open the better, both for the discoverer(s) and for the field.

Tell the truth

Telling the truth about data is an absolute requirement in science. Apart from being the ultimate betrayal of scientific ethics, a lie can set everyone on the wrong course—including the perpetrator. The public revelation of such deception is likely to destroy a career. Those who withhold credit may get away with it for a time, but they run a risk if they want to be recognised ultimately as luminaries in their field.

Be generous and culturally aware

Freely acknowledging the achievements of others is a sure sign of someone who is confident of their own worth and integrity. Give credit where it is due, and acknowledge the work that came before yours and made your discovery possible. No senior scientist is ever hurt by giving precedence in authorship to junior colleagues who have done much of the hands-on work for a research paper.

Making virulent personal attacks on others in public, especially if the person concerned is young and inexperienced, can ultimately be counter-productive. Even if he or she is guilty of sloppy thinking or poorly presented data, it's easy enough to take someone aside later and talk through their conclusions. If they're obtuse and dogmatic, they will soon disappear from science anyway. Science is largely self-correcting, though this can take a while.

Be aware that some cultures are much more attuned than others to receiving (and returning) very direct and even harsh commentary. The type of intellectual sword play and blood-letting characteristic of an Oxbridge common room or high table can, for instance, seem both arrogant and vicious to those who live in the more polite, but concealed, world of US academia. Remember that, as George Bernard Shaw observed, 'England and America are divided by a common language'. Things may seem more familiar than they actually are. It's also important to know the sub-culture. Australian scientists are generally much more accustomed to tough criticism than the population at large, who may be inclined to punch you on the nose if you're too direct.

When it comes to awards, prizes, election to national academies and the like, it is probably important to have as few enemies as possible. Awards are neither gifts at the disposal of the gods on Mount Olympus nor the inevitable outcome of some automatic process of merit recognition.

Selection committees consist of people who are broadly involved in the particular area of interest. At least some committee members will know a lot more about the candidate than the information that appears in the letters of recommendation and the published list of achievements. A strong, convincing and committed opponent can take down almost anyone.

Though it is likely that a truly spectacular discovery or body of achievement will ultimately be recognised, no matter what the personal characteristics of the recipient, there are situations where the decision could go either way. A record of vicious behaviour, or a suggestion that the individual concerned claims undue priority, can seriously damage a case. With the Nobel Prizes, for example, the decision can go to one field or another. There will always be alternative 'tickets'.

Be persistent and tenacious, but be prepared to fail

The old 'Protestant work ethic' is true for science: nothing worthwhile is likely to be easy. If at first you don't succeed, try, try, try again. Sayings like this pretty much describe the scientific life. My 8th-grade teacher, Miss Thompson, drummed into us: 'Good, better best, may you never rest, till your good is better, and your better best'. Maybe Miss T set some up for a life of deep neurosis, but anyone who wants to do experimental science has to be emotionally resilient. Murphy's Law—'Anything that can go wrong will go wrong'—certainly rules in experimental biology. The vital thing is to identify the problem, not apportion blame to one or other member of the research group (unless, of course, it is clearly justified). The Murphy experience, nevertheless, must be a hell of a lot worse for the 'big guys' with something like losing a Mars probe.

Every serious biomedical scientist will have had the experience of 'reverse alchemy', seeing what first looks like gold turn slowly into lead: an apparent breakthrough turns out to be a false trail that just can't be repeated in subsequent experiments. The process of moving systematically from high to low over an interval of weeks to months can be summarised by the Latin, '*sic transit gloria*', thus passes glory. When this happens to you—and it invariably will—see it as a good time to break the cycle, stop the study and go to the pub or on vacation.

People who can't deal with failure, or can't acknowledge to themselves that they have been wrong, should probably avoid a life based in research. Nonetheless, it can be the case that those who are the most creative live on some sort of psychological edge. Such individuals have to develop strategies for dealing with the inevitable downs if they want to work in experimental biology. Sometimes they start well, but just can't continue. Most scientists who've been involved in leading a research effort over the long-term have had to deal with this type of tragedy.

Your time is precious

Winners and high-achievers will tell you that time is your greatest asset. It's accepted that novelists, painters and poets can be precious about protecting their creative time and space, but this territory isn't so clearly defined for those who do research for a living. Scientists work in large organisations and belong to global communities that organise meetings, and national and international societies. These things take up time! They are essential activities, and it is important that such roles should not be left only to those who are at the end of their careers. Even so, it is also necessary to be judicious and to set definite limits. 'Death by committees' is a particular trap for women scientists, who often have great negotiating skills but can be effectively drained by the demands of commitments, which may arise from the need for gender balance on this or that committee. That's fine for someone headed towards a career in academic administration, but such a commitment should be a conscious decision. When those committees come calling, just learn to say no.

Avoid prestigious administrative roles

Those bright people who accept a role as director, dean or president early in a career may well rule themselves out of the top league in the awards game. Leading a major institution is, of course, an energy-consuming activity. The same is true for running a major research program. My personal sense is that smart human beings commit themselves to what they like doing best. Some Nobel Prizes for experimental physics have gone to intellectually incisive, top administrators, but that isn't true for most areas of science. On the other hand, administrators earn the top dollars in the academic hierarchy, and those skills are increasingly in demand.

People who accept such posts sometimes win a Nobel Prize for work done earlier in their careers. Others who make the trip to Stockholm before they're too ancient often go on to be very effective university presidents or directors of prestigious research organisations. However, it's also the case that many who have the creativity and insight to succeed in research lack the types of skills and commitment that go into making an outstanding administrator. As Polonius in *Hamlet* would have it: 'This above all, to thine own self be true'.

Take care of yourself and live a long time

Given the nature of the Peace Prizes, a summons to Oslo is likely to come fairly soon after the achievement that is being recognised, but writers, scientists and economists may need to hang in there. It can take fifty years from the point of making a big discovery to the time that a Nobel committee comes to a decision that, at least from your point of view, is exactly the right one. Good habits start early: eat and drink moderately, take vacations, don't smoke or overuse recreational drugs (alcohol included), take regular exercise, avoid extreme sports, and seek professional help for suicidal thoughts. Scientists are a varied bunch and I know of very talented and effective individuals who have been taken out by each of the above factors. Given a way of life that is often more solitary, creative writers are likely to be even less armoured against such dangers.

Have fun, behave like a winner

See all the above.

On a more serious note, I want to emphasise that, while doing the type of work that leads to Nobel Prizes inevitably has its low points, on the whole it offers immense fulfilment. There can surely be no better feeling than the sense of having achieved an experimental result, or written a novel, poem or scientific paper that is personally satisfying, substantial and accessible, and is out there for the scrutiny of others. Nothing offers greater intellectual excitement than discovering something that no human being can ever have known before. Being able to live in a way that combines work with at least a measure of creativity is an immense privilege. My continued involvement in experimental science reflects this passion for discovery. Only a lunatic would expect to win a second Nobel Prize, so that certainly isn't the motive.

This is ultimately what science is about. Like most scientists I work in big institutions where, increasingly and inevitably, I don't know all the players, especially the younger ones, though they recognise me. If, for instance, I ride up alone in an elevator with some young postdoc whom I may never have seen before, I ask: 'How is it going, and what are you doing?' The invariable experience is that they are delighted to be asked and are just bursting to summarise their particular science story. For those with the right mix of curiosity and commitment, the sense of probing a difficult question, of uncovering some basic—though maybe small—truth gives the greatest possible satisfaction. It isn't for everyone, but for those who get the message this is a good and honest way to live.

Appendixes

Appendixes 1 and 2 are reproductions of the
two short research reports published during 1974 in
Nature that were the basis of the 1996 Nobel prize to
Peter Doherty and Rolf Zinkernagel.
Appendix 3 was printed in the "Hypothesis" section
of *The Lancet*, which was, at the time, one of the few
venues available in the biomedical literature for publishing
theoretical articles in a widely read format.

1

Restriction of in vitro T cell-mediated cytotoxicity in lymphocytic chorio-meningitis within a syngeneic or semiallogeneic system[*]

Recent experiments[1-3] indicate that cooperation between thymus derived lymphocytes (T cells) and antibody-forming cell precursors (B cells) is restricted by the H-2 gene complex. Helper activity *in vivo* operates only when T cells and B cells share at least one set of H-2 antigenic specificities. Evidence is presented here that the interaction of cytotoxic T cells with other somatic cells budding[4-5] lymphocytic choriomeningitis (LCM) virus is similarly restricted.

Both the cytotoxic assay used and the characteristics of the cells involved have been described previously[6-8]. Briefly, monolayers of C3H mouse fibroblasts (L cells) are grown in plastic tissue culture trays, infected with a high multiplicity of the WE3 strain of LCM virus and the cells labelled with ^{51}Cr and overlaid (40:1) with the spleen cell preparation to be tested. Supernatants are removed between 15 and 16 h later and % ^{51}Cr release calculated[7]. Results are expressed as mean ± s.e.m. for four replicates. Cytolysis of infected

[*] Reproduced with permission from *Nature*, vol. 248, No. 5450, pp. 701–702, April 19, 1974.

L cells by CBA/H immune spleen cells has been shown to be a property of specifically sensitised thymus-derived lymphocytes, which act in the absence of both macrophages and substances secreted into the medium at large[6–8].

Various strains of mice were injected intracerebrally (i.c.) with 300 mouse LD_{50} of WE3 LCM virus. Mice were sampled 7 d later when, in CBA/H mice, maximal cytotoxic activity is found in lymphoid tissues[6–7]. Only

Table 1 Cytotoxic activity of spleen cells from various strains of mice injected i.c. 7 d previously with 300 LD_{50}* of WE3 LCM virus for monolayers of LCM-infected or normal C3H ($H-2^k$) mouse L cells.

Experiment	Mouse strain	H-2 type	% ^{51}Cr release[†] Infected	Normal
1	CBA/H	k	65.1 ± 3.3	17.2 ± 0.7
	Balb/C	d	17.9 ± 0.9	17.2 ± 0.6
	C57Bl	b	22.7 ± 1.4	19.8 ± 0.9
	CBA/H × C57Bl	k/b	56.1 ± 0.5	16.7 ± 0.3
	C57Bl × Balb/C	b/d	24.8 ± 2.4	19.8 ± 0.9
	nu/+ or +/+		42.8 ± 2.0	21.9 ± 0.7
	nu/nu		23.3 ± 0.6	20.0 ± 1.4
2	CBA/H	k	85.5 ± 3.1	20.9 ± 1.2
	AKR	k	71.2 ± 1.6	18.6 ± 1.2
	DBA/2	d	24.5 ± 1.2	21.7 ± 1.7
3	CBA/H	k	77.9 ± 2.7	25.7 ± 1.3
	C3H/HeJ	k	77.8 ± 0.8	24.5 ± 1.5

* Other mice were injected with 2×10^6 LD_{50}, but levels of specific release were invariably lower due to the high dose immune paralysis[8,20] associated with viscerotropic (WE3) LCM virus.

† % ^{51}Cr release by normal spleen cells on infected targets ranged from: (experiment 1) 17.1 ± 0.3 to 20.0 ± 0.7; (experiment 2) 20.0 ± 1.4 to 25.3 ± 0.7; (experiment 3) 27.2 ± 2.0.

spleen preparations[6] from mice sharing at least one set of H-2 antigenic specificities with the target monolayer caused high levels (40% to 50%) of specific lysis (Table 1). Spleen cells from nude control (nu/+ or +/+) mice, derived locally from CBA and Balb/C stock (Dr J. B. Smith, personal communication), were less active and lymphocytes from histoincompatible mice caused minimal specific release (< 5%) of ^{51}Cr.

Spleen preparations from mice immunised 10 and 13 d previously were also assayed, as Marker and Volkert[9] have reported that maximal cytotoxicity of C3H lymphocytes for L cells infected with the Traub strain of LCM virus occurs at 11 d after inoculation. High levels of specific ^{51}Cr release were again recognised only in the histocompatible system (Table 2) activity declining, as has been shown previously[7], from a peak on day 7.

Demonstration of reciprocal exclusion of cytolysis was essential to establish that mice possessing other than H-2^k antigenic specificities are capable of generating cytotoxic

Table 2 % ^{51}Cr release* from infected C3H L cells overlaid with spleen cells from mice sampled at 7, 10 and 13 d after intravenous inoculation with 2,000 LD$_{50}$ of WE3 LCM virus.

Mouse strain	Days after inoculation		
	7	10	13
CBA/H	72.0 ± 2.0	66.4 ± 1.4	27.5 ± 0.5
Balb /C	26.1 ± 0.7	28.0 ± 1.6	22.7 ± 1.8
C57Bl	27.3 ± 1.1	24.3 ± 1.8	24.0 ± 0.4

* Levels of ^{51}Cr release due to overlaying normal L cells with immune spleen cells, infected L cells with control spleen cells or with medium alone ranged from 17.1 ± 0.4 to 24.0 ± 1.4. Other mice were injected with 2×10^6 LD$_{50}$, but levels of specific release were invariably lower.

T cells. Comparisons were thus made using similar targets from allogeneic mouse strains. Peritoneal macrophages were obtained[10] from normal Balb/C and CBA/H mice, cultured in plastic tissue culture trays and infected[11] with WE3 LCM virus. Specific lysis was restricted to the syngeneic system (Table 3). Comparable levels of specific ^{51}Cr release from isologous infected macrophages were caused by Balb/C and CBA/H spleen cells (overlaid at 20:1) from mice infected at the same time with the same dose of LCM virus, whereas histoincompatible macrophages were

Table 3 % ^{51}Cr release from normal and infected peritoneal macrophages by spleen cells from control mice and from mice injected i.c. with 300 LD$_{50}$ of WE3 LCM virus 7 d previously.

| | | % ^{51}Cr release from macrophages | | | |
| | Macrophage | Experiment 1 | | Experiment 2 | |
Spleen cells	source	Infected	Normal	Infected	Normal
Balb/C Immune	Balb/C	61.8 ± 4.2^c	27.6 ± 1.9^e	77.5 ± 4.2^d	47.0 ± 3.5^d
Anti-θ*		ND	ND	40.6 ± 2.5^e	ND
N ascitic*		ND	ND	90.0 ± 2.7	ND
Control		42.0 ± 4.8^a	40.5 ± 5.2^a	49.6 ± 2.5	43.5 ± 1.6
CBA/H Immune		42.7 ± 6.7^a	33.7 ± 5.4^a	32.9 ± 3.0^a	48.6 ± 3.9^a
Control		28.0 ± 4.1	40.5 ± 5.2	46.5 ± 3.7	39.7 ± 4.3
CBA/H Immune	CBA/H	69.1 ± 2.8^c	30.9 ± 3.4^c	72.5 ± 5.2^d	40.0 ± 2.9^d
Anti-θ		ND	ND	44.0 ± 2.5^d	ND
N ascitic		ND	ND	74.3 ± 8.4	ND
Control		34.2 ± 1.1	35.1 ± 3.7	46.5 ± 3.6	44.4 ± 6.2
Balb/C Immune		46.2 ± 3.3^a	30.4 ± 3.8^b	44.0 ± 2.9^a	41.0 ± 2.4^a
Control		34.9 ± 5.7	33.7 ± 5.6	40.5 ± 2.5	41.0 ± 2.4

* Treated with AKR anti-θ (C3H) ascitic fluid and guinea pig complement, or normal AKR ascitic fluid and guinea pig complement. a, b, c, d, e, Differences by Student's t test between values for immune spleen cells treated with anti-θ ascitic fluid or normal ascitic fluid, immune and control spleen cells overlaid on infected macrophages (infected column), or immune spleen cells overlaid on infected and normal macrophages (normal column). a, $P > 0.05$; b, $P < 0.05$; c, $P < 0.02$; d, $P < 0.01$; e, $P < 0.001$. ND, Not done.

not damaged. Lysis was completely abrogated by treatment with AKR anti-θ ascitic fluid and guinea pig complement, but not by normal AKR ascitic fluid and complement. Though levels of ^{51}Cr release from macrophages were more variable than for L cells, probably because of inconsistencies in target cell concentrations and the higher background of non-specific lysis, the effect was both highly significant and repeatable.

The ability of T cells to cause lysis across an allogeneic barrier, when lymphocytes are sensitised to antigens specified by the H-2 gene complex, is well established[12,13]. This also applies to L cells (H-2^k), which are readily lysed by spleen cells from C57Bl (H-2^b) mice stimulated in mixed lymphocyte culture by CBA/H (H-2^k) but not by Balb/C (H-2^d) lymph node cells. Sufficiently close association for lysis to occur is possible if the T cells is sensitised to allo-antigens present on the surface of the target. Interaction between immune lymphocytes and cells expressing antigens expressed by LCM virus is, however, apparently confined to a histocompatible system, perhaps because it is only in this situation that the necessary intimacy of contact is achieved.

This restriction may possibly be overcome by previous treatment of the target population with trypsin. Balb/C (H-2^d) immune spleen cells will lyse recently trypsinised L cells (H-2^k) infected with the E-350 strain of LCM virus, the effect being completely abrogated by treatment with a rabbit anti-mouse brain serum cytotoxic for T cells[14]. Our experiments with previously trypsinised WE3-infected L cells have, to date, given equivocal results because of the high background of non-specific ^{51}Cr release.

An alternative possibility that must be considered in LCM is that the process of virus maturation[5-6] through the cell membrane causes changes in self components, which are recognised only within the syngeneic or semi-allogeneic system. There is ample evidence[15-16], for instance, that concentrations of H-2 antigens in the cell membrane are decreased in cells productively infected with budding viruses. The cytotoxic T cell may thus be recognising altered self, the implication being that LCM is essentially an auto-immune phenomenon.

These results impose a possible constraint on attempts to demonstrate cytotoxic T cell in infections of man and domestic mammals, where histocompatible cell lines and inbred strains are not available. Perhaps isologous macrophages or lectin-transformed peripheral blood leukocytes may prove suitable targets in at least some disease states. Restriction of cell-mediated cytotoxicity within a syngeneic or semiallo-geneic system may prove a reliable index of T cell involvement in species where the θ marker is not available, lysis across this barrier indicating an antibody-associated process[17-19].

Dr Zinkernagel is supported by the Schweizerische Stiftung Fuer Biologisch-Medizinische Stipendien.

R. M. Zinkernagel
P. C. Doherty

Department of Microbiology,
John Curtin School of Medical Research,
Australian National University,
Canberra

Received December 10, 1973.

Appendix 1

[1] Kindred, B., and Shreffler, D. C., *J. immun.*, **109**, 940 (1972).

[2] Katz, D. H., Hamaoka, T., and Benacerraf, B., *J. exp. Med.*, **137**, 1405 (1973).

[3] Katz, D. H., Hamaoka, T., Dorf, M. E., and Benacerraf, B., *Proc. natn. Acad. Sci., U.S.A.*, **70**, 2624 (1973).

[4] Abelson, H. T., Smith, G. H., Hoffman, H. A., and Rowe, W. P., *J. natn. Cancer Inst.*, **42**, 497 (1969).

[5] Kajima, M., and Majde, J., *Naturwissenschaften*, **57**, 93 (1970).

[6] Zinkernagel, R. M. and Doherty, P. C., *J. exp. Med.*, **138**, 1266 (1973).

[7] Doherty, P. C., Zinkernagel, R. M., and Ramshaw, I. A., *J. Immun.* (in the press).

[8] Doherty, P. C., and Zinkernagel, R. M., *Transpln. Rev.*, **18** (in the press).

[9] Marker, O., and Volkert, M., *J. exp. Med.*, **137**, 1511 (1973).

[10] Blanden, R. V., Mackaness, G. B., and Collins, F. M., *J. exp. Med.*, **124**, 585 (1966).

[11] Mims, C. A., and Subrahmanyan, T. P., *J. Path. Bact.*, **91**, 403 (1966).

[12] Cerottini, J.-C., Nordin, A. A., and Brunner, K. T., *Nature*, **228**, 1308, (1970).

[13] Cerottini, J.-C., and Brunner, K. T., *Adv. Immun.* (in the press).

[14] Cole, G. A., Prendergast, R. A., and Henney, C. S., *Fedn. Proc.*, **32**, 964 (1973).

[15] Hecht, T. T., and Summers, D. F., *J. Virol.*, **10**, 578 (1972).

[16] Lengerova, A., *Adv. Cancer Res.*, **16**, 235 (1972).

[17] Perlmann, P., Perlmann, H., and Wigzell, H., *Transpln. Rev.*, **13**, 91 (1972).

[18] MacLennan, I. C. M., *Transpln. Rev.*, **13**, 67 (1972).

[19] Steele, R. W., Hensen, S. A., Vincent. M. M., Fuccillo, D. A., and Bellanti, J. A., *J. Immun.*, **110**, 1502 (1973).

[20] Hotchin, J., *Monogr. Virol.*, **3**, 1 (1971).

2

Immunological surveillance against altered self components by sensitised T lymphocytes in lymphocytic choriomeningitis*

The cytotoxic activity[1,2] of immune thymus-derived lymphocytes (T cells) for ^{51}Cr-labelled fibroblasts, or macrophages infected with lymphocytic choriomeningitis (LCM) virus is restricted by the H-2 gene complex[3,4]. Specific lysis of LCM-infected monolayer cultures occurs only when targets and overlaying, sensitised T cells share at least one set of H-2 antigenic specificities.

Operationally, this restriction may reflect one of two quite distinct mechanisms[3,4]. First, H-2 compatibility is essential for sufficiently close association[3] between lymphocyte and target cell for lysis to occur. This intimacy model implies either that the H-2 gene complex specifies a product, or products, involved in self-recognition, or that there is some mutual interaction between H-2 antigens. Such a process would be additional to recognition of viral antigen by the T cell receptor. The alternative possibility is that infection with LCM virus modifies self components in a way recognised only within a H-2 compatible system.

* Reproduced with permission from *Nature*, vol. 251, No. 5475, pp. 547–48, October 11, 1974.

Altered self may be thought of as changes in H-2 antigens (or in structures coded for in the H-2 region) induced by the process of virus synthesis, or as some complex of viral and H-2 antigens.

Lymphocytes from F_1 immune mice lyse virus-infected targets of parent H-2 type as effectively as do syngeneic T cells[1]. If the intimacy model is correct, therefore, F_1 mice need possess only one clone of sensitised T cells, recognising viral antigen by a specific receptor and like H-2 antigen, of either parent type, in some non-immunological way (Fig. 1). The altered self hypothesis, however, requires the presence of at least two clones of T cells, each reactive to modified H-2 of one parent type.

These possibilities have been examined using the following experimental system. When LCM-immune T cells are injected into immunosuppressed[5], virus-infected recipients they home equally well to lymphoid tissue of syngeneic or allogeneic hosts, but continue to multiply only in the syngeneic system[6]. Further replication is apparently not triggered by free virus, but is dependent on thymus-derived lymphocytes being exposed to histocompatible, virus-infected target cells.

Continued proliferation of F_1 ($H-2^{k/b}$) immune spleen cells with capacity to lyse $H-2^k$ LCM-infected monolayers occurs only in recipients with $H-2^k$ antigenic specificities, and not in those possessing $H-2^d$ (Table 1). These sensitised T cells are of donor origin, as lymphocytes recovered from H-2 compatible AKR recipients adoptively immunised with CBA/H × C57Bl spleen cells were shown to bear θ C3H (Table 2). The intimacy hypothesis, therefore, is probably not correct, as the postulated single clone of F_1 lymphocytes (Fig. 1) should continue to proliferate equally

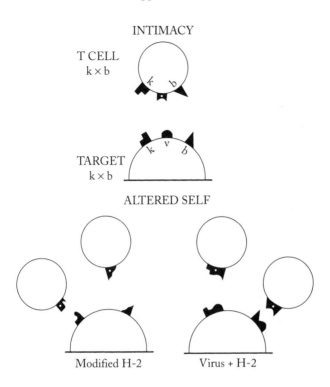

INTIMACY

T CELL
k × b

TARGET
k × b

ALTERED SELF

Modified H-2   Virus + H-2

Fig. 1 *Capacity of sensitised F_1 (H-2$^{k/b}$) T cells to interact only with histocompatible virus–infected target cells may be considered to reflect any one of the models shown. The intimacy concept proposes a single immunologically specific T cell receptor for viral (v) antigen, additional to a requirement for physiological interaction coded for by the H-2 gene complex (mutuality between either H-2^k or H-2^b). The two models proposed for altered self postulate that, in each case, there are at least two T cell populations with receptors of different immunological specificities recognising modified H-2, or virus + H-2 of either parent type.*

Table 1 Cytotoxic activity of donor T cells in LCM-infected irradiated recipients*

Donor cells*	Recipient strain	H-2 type	% ^{51}Cr release[†] from L cells (H-2^k) Infected	Normal
Experiment 1				
Immune	CBA/H × BALB/c	k/d	51.2 ± 0.5	12.5 ± 1.9
	C57Bl × BALB/c	b/d	17.3 ± 0.1	13.9 ± 1.3
	BALB/c	d	16.1 ± 1.8	13.0 ± 1.3
Normal	BALB/c	d	14.4 ± 1.9	11.7 ± 1.1
Experiment 2				
Immune	CBA/H	k	83.7 ± 5.4	19.0 ± 1.4
	C57Bl	b	23.3 ± 1.3	19.1 ± 1.0
Normal	C57Bl	b	20.3 ± 1.2	18.5 ± 1.7

* Donor CBA/H × C57Bl F$_1$ were injected intracerebrally with 300 LD50 of WE3 LCM virus and killed when clinically affected 7 d later. Recipients were irradiated (850 r.) 24 h before i.v. injection of 10^6 LD50 of WE3 LCM virus, and inoculated i.v. with $5.0 × 10^7$ spleen and lymph node cells 6 h later.

[†] % ^{51}Cr release above normal values is a measure of activity of H-2 compatible sensitised T cells[1–4]. Spleen cells from LCM-immune mice caused % ^{51}Cr release of 46.6 ± 0.9% from infected L cells and 15.2 ± 1.3% from normal L cells (experiment 1).

Table 2 Evidence that sensitised T cells are of donor origin

Donor	Recipient	% ^{51}Cr release from L cells Treatment	Infected	Normal
CBA/H × C57Bl F$_1$	AKR	Anti-θ + C	22.3 ± 1.2	21.4 ± 2.5
H-2$^{k/b}$ θC3H	H-2^k, θAKR	N ascitic + C	89.1 ± 1.8	23.3 ± 1.2

The protocol is the same as in Table 1, except that a proportion of lymphoid cells were treated[3] with AKR anti-θ (C3H) ascitic fluid and rabbit complement or normal AKR ascitic fluid and complement.

well in recipients of either parent strain. The results cannot, however, exclude this model totally, the reservation being that there may be alleleic exclusion in the F$_1$ of this hypothetical self-recognition product. The simplest explanation of our results is, however, that there are sensitised T cells of at least two specificities in LCM-infected H-2$^{k/b}$ mice, each recognising a complex of virus + H-2 (or modified H-2) of one parent type. Recirculating T cells may function essentially to survey the integtity of transplantation antigens, or structures coded for by the H-2 gene complex. Recognition of cell surface changes due to virus infection, chemical modification[7] or genetic difference (alloantigens) may then be accommodated within the same model.

We thank Professor G. L. Ada, Dr A. J. Cunningham, Dr R. V. Blanden and Andrew Hapel for discussion and particularly Dr L. Pilarski for emphasising the possibility of a complex of virus-induced and H-2 antigen. The anti-θ serum was a gift of Dr A. J. Cunningham.

Rolf M. Zinkernagel
Peter C. Doherty

Department of Microbiology,
The John Curtin School of Medical Research,
Canberra, A.C.T., Australia

Received June 24; revised August 19, 1974.

[1] Zinkernagel, R. M., and Doherty, P. C., *Scand. J. Immun.* (in press).
[2] Doherty, P. C., Zinkernagel, R. M., and Ramshaw, I. A., *J. Immun.*, **112**, 1548–1552 (1974).
[3] Zinkernagel, R. M., and Doherty, P. C., *Nature*, **248**, 701–702 (1974).
[4] Doherty, P. C., and Zinkernagel, R. M., *Transplantn Rev.*, **19**, 89–120 (1974).
[5] Gilden, D. H., Cole, G. A., Monjan, A. A., and Nathanson, N., *J. exp. Med.*, **135**, 860–873 (1972).
[6] Doherty, P. C., and Zinkernagel, R. M., *J. Immun.* (in press).
[7] Shearer, G. M., *Eur. J. Immun.* (in press).

3

A Biological Role for the Major Histocompatibility Antigens*

P. C. Doherty
R. M. Zinkernagel

Department of Microbiology,
The John Curtin School of Medical Research,
Canberra, A.C.T., Australia

Summary

A central function of the major histocompatibility (H) antigens may be to signal changes in self to the immune system. Virus-induced modification of strong transplantation antigens apparently results in recognition by thymus-derived lymphocytes (T cells), with subsequent clonal expansion and immune elimination of cells bearing non-self determinants. The extreme genetic polymorphism found in the major H antigen systems of higher vertebrates may reflect evolutionary pressure exerted by this immunological surveillance mechanism.

* Reprinted from *The Lancet* (June 28, 1975): 1406.

Introduction

The present hypothesis is derived from experimental evidence that immune T cells in three of the most prevalent naturally occurring virus diseases of mice—lymphocytic choriomeningitis[1] (L.C.M.), ectromelia[2] (mouse pox), and paramyxovirus[3] (Sendai) infection—are apparently sensitised to altered self antigens. Genetic mapping studies have established that the self components involved are specified at, or near to, either of the two loci coding for the major histocompatibility (H) antigens.[2-5] Monitoring of H antigen structural integrity by recirculating thymus-derived lymphocytes may thus be an essential feature of immunological surveillance.[6] Furthermore, existence of such a mechanism provides a biological basis for the evolution of strong transplantation antigen systems, a physiological role for which has been conspicuously lacking.[7]

The H-2 Gene Complex

The majority of this discussion is concerned with H-2, as our evidence is derived solely from mouse experiments. However, H-2 and HL-A are comparable systems,[8] and the argument may apply generally to higher vertebrates. Before proceeding further it is necessary to briefly review the nature of H-2.[9,10] The major transplantation antigens of mice are coded for by the H-2 gene complex, situated on chromosome 17 (linkage group IX). The complex is generally inherited as a single unit, and is divided conventionally into four regions: K, I, S, and D. Individual H-2 types are described by the use of small letters—e.g., H-2^k (i.e., H-2K^k-D^k). Genes specifying the major H antigens are sited at H-2K and H-2D, which behave as single loci with multiple alleles and are equivalent to the LA and FOUR loci of man. The products of these two regions are apparently independent on the cell surface, as they do not co-cap when exposed to highly specific antisera.[11] Components of both H-2 and HL-A antigens recognised by antibody are polypeptides, thought to be organised in a two-unit structure similar to the constant region domains of immunoglobulins.[12,13] Serologically determined H-2 private specificities (strong transplantation antigens) are associated only with a single H-2 type, whereas H-2 public specificities may be shared between a number of H-2 types. The I region in mice is concerned with regulation of some antibody responses requiring T cell help,[14] and S is associated with a serum-protein which is a useful marker for defining genetic recombination events.

Requirement for H-2 Comparibility

Specific cell-mediated lysis of ^{51}Cr-labelled fibroblasts, macrophages, or tumour cells occurs only if immune donors of T cells and virus-infected targets are compatible at either H-2K or H-2D[1-5,15] Successful interaction apparently reflects mutuality of serologically defined H-2 private antigenic specificities, though this may simply be a marker for closely linked genes coding for differences not recognised by circulatory antibody. Identify at the I-S region is neither sufficient nor necessary for lysis to occur. Differences at the M locus, which is involved in mixed lymphocyte reactions and is sited on another mouse chromosome,[16] are also irrelevant. Less detailed animal experiments have shown that the same H-2 compatibility requirement applies also to T cell effector activity in vivo.[17,18] Furthermore, evidence from other laboratories indicates that the T cell response to Rous sarcoma virus in chickens,[19] and to trinitrophenyl (T.N.P.)-modified mouse lymphocytes,[20-21] is similarly restricted.

There are several possible explanations for this phenomenon. Any argument based on inhibition due to presence of alloantigens[22] (different H-2 antigens) has been eliminated by cell-mixture experiments and by the use of various parent-F1 combinations both in vitro and in vivo.[5,17] This leaves two major alternatives, which we have designated *intimacy* or *altered self*.[23]

The *intimacy* model may be summarised as follows: the H-2 gene complex codes for a physiological self-recognition mechanism,[24-26] which operates in addition to immunologically specific recognition of viral antigen by the T cell receptor. This dual interaction hypothesis implies that the host need generate T cells of only one broad specificity reactive to virus. The H-2 compatibility requirement would thus reflect operation of the physiological self-recognition system. There are two immediately apparent problems with this idea. First, the fact that compatibility at either H-2K or H-2D is sufficient for lysis of virus-infected cells to occur implies that there must be two separate physiological interaction mechanisms, coded for at H-2K or H-2D. Second, it does not explain sensitisation to alloantigens, the cell-mediated immune response to which is essentially similar to that resulting from infection of syngeneic (self) cells with virus. Obviously, allogeneic cells must lack this self-recognition system.

The *altered self* concept is simpler, requiring that a single (per clone) immunologically specific T cell receptor recognises virus-induced changes of syngeneic H antigens —i.e., different T cell clones from H-2K^k-D^k L.C.M.-infected mice are sensitised to altered self antigens coded for by H-2K^k or H-2D^k, and do not include specificities reactive to virus-induced changes in self coded for by H-2K^d or H-2D^d as these antigens are not encountered during immunisation.

Intimacy of Altered Self

Experiments aimed at differentiating between these two possibilities have been based on two distinct protocols; (1) competitive inhibition of cytotoxicity in vitro by addition of "cold", unlabelled target cells, and (2) further multiplication of immune T cells transferred to irradiated, virus-infected recipients. These experiments have been described elsewhere,[5,21,27] and will not be discussed here. The overall conclusion from both approaches is that different clones of sensitised T cells recognise self components specified at, or near to, H-2K or H-2D.

The *intimacy* model can be accommodated with these findings in either of two ways:

(1) Clones of T cells express only one of the postulated physiological interaction structures coded for at H-2K or H-2D, and in heterozygotes there is superimposed allelic exclusion at these two loci. This complex form of genetic exclusion is, to our knowledge, extremely unlikely.

(2) The self components involved in mutual interaction form an integral part of both the T cell receptor and of the antigen, and this receptor complex is clonally expressed. This second possibility is operationally identical with the *altered self* concept: structures coded for at

H-2K or H-2D must form at least part of the antigen recognised by the T cell. Reactivity to alloantigens, which may be regarded as slightly different from self, may be similarly divided with respect to these two regions.[28]

Virus-induced modification of self may reflect some complex of viral and H antigens, expression of H antigens that are not normally seen on the cell surface (e.g., neoantigens physically exposed during virus budding or products of derepressed host genes[29]). At this stage we do not understand the biochemical nature of alteration in self, nor is such knowledge central to the present argument.

Altered Self and Immunological Surveillance

The central implication of the *altered self* concept is that cell-mediated immunity to virally or chemically modified cells and to allogeneic cells[30] is of the same order. Reactivity to xenoantigens (major H-antigens from a different species) may also be accommodated within this model, though differences from self will obviously be greater and the T cell response is concurrently much weaker.[31,32] Monitoring of H antigen structural integrity by recirculating T cells may thus be an essential feature of immunological surveillance, with specific recognition leading to clonal expansion and elimination of cells bearing major H antigens seen as not-self. Selective pressure exerted by this immunological surveillance mechanism may have been of great importance in the development of the immune system in general, perhaps along the lines suggested by Jerne.[33]

Reactivity associated with strong transplantation antigens may not be characteristic of all classes of T cell response. It would obviously be disadvantageous for surveillance T cells to kill all macrophages and B cells (antibody-forming cell precursors) expressing non-self antigenic determinants. This may tend to happen in conditions such as L.C.M., where the virus infects many cells in lymphoid tissue[34] and cell-mediated immunity is, at least initially, more prominent than the humoral response.[35,36] In the main, however, interaction between T cells and B cells may reflect the evolution of a functionally separate class of helper T cells,[37] programmed to recognise antigen-induced modification of structures (Ia) coded for by genes in the I region.[38-40] High concentrations of Ia are found on B cells and macrophages.[41] Modification of Ia may thus lead to T cell helper activity and to a more effective antibody response, whereas changes in H antigen structure result in immune elimination mediated by surveillance T cells. Regulation[42] of the overall immune response may reflect complex interactions between these two systems.[2,21,40]

Evolutionary success of an immunological surveillance mechanism associated with a particular set of H antigens may, in turn, have led to strong expression of these antigens. Major H antigen systems such as H-2 and HL-A are recognised only in higher vertebrates and have not been defined in life forms more primitive than amphibia,[43] though they may also exist in the bony fishes.[44] Also, the tissue, distribution of major H antigens may reflect the extent of surveillance T cell activity. Macrophages and lymphoid cells are particularly exposed to infectious agents and express high concentrations of H-2.[45] In the brain, however, where implanted skin allografts do not elicit an immune response,[46] expression of H-2 antigens is minimal.[45]

The Basis of H-Antigen Polymorphism

The major H antigen systems of mice and men show extreme genetic polymorphism equalled, in vertebrates, only by that associated with the immunoglobulins.[8,47,48] No generally acceptable reason has been advanced for this profound variability, maintenance of which is best explained by there being selective advantage for heterozygosity in the absence of overall evolutionary pressure favouring any particular H antigen type.[47]

We wish to propose two separate lines of explanation for this extreme variability in H antigens. Both are based on the realisation that different T cell specificities are associated with self antigens coded for at each of the two major H antigen loci, and may be summarised as follows:

(1) The range of T cell responsiveness is, on a purely mechanistic basis, greater in individual F1 than in homozygous mice.

(2) Some pathogens (or oncogenic processes) may not cause an immunogenic modification of self associated with a particular H allele, and existence of multiple polymorphism thus minimises the risk of there being general unresponsiveness throughout the population. This latter possibility has been suggested independently by Shearer and his colleagues.[21]

Both arguments may also be used to explain the evolutionary advantage of there being two, rather than one, H antigen loci.

Homozygous L.C.M.-immune mice apparently possess T cells of two broad specificities, sensitised to modification of H antigens coded for at H-2K or H-2D.[5,27] Heterozygotes would thus generate T cells of four specificities, associated with H-2K and H-2D of each parental haplotype. The range of T cell responsiveness is thus effectively doubled in the F1. Does this result in enhanced T cell effector function in mice heterozygous at the H-2 gene complex?

Analysis of data from in-vitro cytotoxic assays reveals that L.C.M.-immune F1 T cells lyse virus-infected targets of one parental H-2 type almost as effectively (by various estimates from 70% to 100%) as do fully syngeneic parent-strain effector lymphocytes.[5] The total response in the F1 associated with both parental H-2 types, which should be equally expressed on the cell surface, could thus be from 140% to 200% of that occurring in parental homozygotes. Evidence from in-vivo experiments also indicates that T cell-mediated immunopathology is more severe in heterozygous mice infected with L.C.M. virus.[49] The acute pathological process characteristic of L.C.M.[23] is, however, atypical of T cell responses in general. In most virus diseases cell-mediated immunity operates to eliminate infected cells, and consequently to protect the host.[50] Enhancement of T cell reactivity in the F1 would, therefore, generally be of selective advantage and would tend to maintain an overall balanced polymorphism in H antigen types.

Superimposed on this mechanistic argument for selective advantage of heterozygotes is the concept, emerging from current experiments, that viruses may not induce immunogenic modifications of all H antigen types.[2,38] In L.C.M. infection the T cell response associated with H-2D appears stronger than that at H-2K. The same may be true for the response to some oncogenic viruses, where resistance is related to H-2D.[51,52] Reactivity to L.C.M. virus associated with H-2K^d may, at least in one mouse strain, be particularly weak, whereas in ectromelia H-2D^k is associated with a minimal response.[2] This raises the possibility that the correlation between susceptibility to a particular infectious disease or oncogenic process and HL-A[53] or H-2 type may reflect absence of an immunogenic modification in major H antigens, rather than operation of controlling immune response (Ir) genes.[14]

Overall selection against a particular H-2 type might occur if only one disease process were prevalent. Most animal population are, however, subject to a great variety of infections. Maintenance of as wide a variety of HL-A or H-2 types would thus be of general advantage, minimising the chance that a particular virus would fail to produce an immunogenic H-antigen modification throughout the population. Again, the net result would tend towards a balanced polymorphism.

[1]   Zinkernagel, R. M., Doherty, P. C. *Nature,* 1974, **248**, 701.
[2]   Blanden, R.V., Doherty, P. C., Dunlop, M. B. C., Gardner, I., Zinkernagel, R. M., David, C. S. *ibid.* 1975, **254**, 269.
[3]   Doherty, P. C., Zinkernagel, R. M. Unpublished.
[4]   Doherty, P. C., Zinkernagel, R. M. *J. exp. Med.* 1975, **141**, 502.
[5]   Zinkernagel, R. M., Doherty, P. C. *ibid.* (in the press).
[6]   Burnet, F. M. *Immunological surveillance.* Sydney, 1970.
[7]   Snell, G. D. *Immunogenetics,* 1975, **1**, 1.
[8]   Amos, D. B. et al. *Fedn. Proc.* 1972, **31**, 1087
[9]   Klein, J., Shreffler, D. C. *Transplant. Rev.* 1971, **6**, 3.
[10] Shreffler, D. C., David, C. S. *Adv. Immun.* 1974 (in the press).

[11] Neauport-Sautes, C., Lilly, F., Silvestre, D., Kourilsky, F. M. *J. exp. Med.* 1973, **137**, 511.

[12] Tanigaki, N., Pressman, D. *Transplant. Rev. 1974, 21*, 15.

[13] Strominger, J. L., Cressell, P., Grey, H., Humphreys, R. E., Mann, D., McCune, J., Parham, P., Robb, R., Sanderson, A. R., Springer, T. A., Terhorst, C., Turner, M. J. *ibid* p. 126.

[14] Green, I. *Immunogenetics*, 1975, **1**, 4.

[15] Gardner, I., Bowern, N. A., Blanden, R. V. *Eur. J. Immun.* (in the press).

[16] Festenstein, H. *Transplant. Rev.* 1973, **15**, 62.

[17] Doherty, P. C., Zinkernagel, R. M. *J. Immun.* 1975, **114**, 30.

[18] Blanden, R. V., Bowern, N. A., Pang, T. E., Gardner, I., Parish, C. R. *Aust. J exp. Biol. med. Sci.* (in the press).

[19] Wainberg, M. A., Markson, Y., Weiss, D. W., Doljanski, F. *Proc. natn. Acad. Sci. U.S.A.* 1974, **71**, 3565.

[20] Shearer, G.M. *Eur. J. Immun.* 1974, **4**, 257.

[21] Shearer, G.M., Rehn, G.R., Garbarino, C.A. *J. exp. Med.* (in the press).

[22] Hellstrom, K. E., Möller, G. *Progr. Allergy*, 1965, **9**, 158.

[23] Doherty, P. C., Zinkernagel, R. M. *Transplant. Rev.* 1974, **19**, 89.

[24] Katz, D. H., Hamaoka, T., Dorf, M. E., Benacerraf, B. *Proc. natn. Acad. Sci. U.S.A.* 1974, **70**, 2624.

[25] Shevach, E. M., Green, I., Paul, W. E. *J. exp. Med.* 1974, **139**, 679.

[26] Rosenthal, A. S., Shevach, E. M. *ibid.* 1973, **138**, 1194.

[27] Zinkernagel, R. M., Doherty, P. C. *Nature*, 1974, **251**, 547.

[28] Brondz, B. D., Egorov, I. K,. Drizlikh. G. I. *J. exp. Med.* 1975, **141**, 11.

[29] Martin, W. J. *Cell Immun.* 1975, **15**, 1.

[30] Cerottini, J.-C., Brunner, K. T. *Adv. Immun.* 1974, **19**, 67.

[31] Lindahl, K. F., Bach, F. H. *Nature*, 1975, **254**, 607.

[32] Wekerle, H., Eshhar, Z., Lonai, P. Feldman, M., *Proc. natn. Acad. Sci. U.S.A.* 1975, **72**, 1147.

[33] Jerne, N. K. *Eur J. Immun.* 1971, **1**, 1.

[34] Mims, C. A. Tosolini, F. A., *Br. J. exp. Path.* 1969, **50**, 584.

[35] Hotchin, J. *Monogr. Virol.* 1971, **3**, 1.

[36] Lehmann-Grube, F. *Virol. Monogr.* **10**, 1.

[37] Wernet, D., Lilly, F. *J. exp. Med.* 1971, **141**, 573.

[38] Kantor, H., Boyse, E. A. *ibid.* (in the press).

39 Erb, P., Feldman, M. *ibid.* (in the press).
40 Zinkernagel, R. M., Doherty, P. C., Blanden, R. V. Unpublished.
41 Hammerling, G. J., Mauve, G., Goldberg, E., McDevitt, H. O. *Immunogenetics*, 1975, **1**, 428.
42 Bretscher, P.A. *Cell. Immun.* 1973, **6**, 1.
43 Du Pasquier, L., Chardonnens, X., Miggiano, V. C. *Immunogenetics*, 1975, **1**, 482.
44 Hildemann, W. H. *in Transplantation Antigens: Markers of Biological Individuality* (edited by B.D. Kahan and R.M. Reisfeld); p. 3. New York, 1972.
45 Edidin, M. *ibid.* p. 95.
46 Medawar, P. B. *Br. J. exp. Path.* 1948, **19**, 58.
47 Bodmer, W. F. *Nature*, 1972, **237**, 139.
48 Burnet, F. M. *ibid.* 1973, **245**, 139.
49 Doherty, P. C., Zinkernagel, R. M. Unpublished.
50 Blanden, R. V. *Transplant. Rev.* 1974, **19**, 56.
51 Muhlbock, O., Dux, A. *J. natn. Cancer Inst.* 1974, **53**, 993.
52 Chesebro, B., Wehrly, K., Stimpfling, J. *J. exp. Med.* 1974, **140**, 1457.
53 Vladutiu, A. O., Rose, N. R. *Immunogenetics*, 1974, **1**, 305.

Abbreviations

AID	Agency for International Development
ANU	Australian National University, Canberra
ARC	Australia Research Council, which supports research in science and the humanities
Big Pharma	Large pharmaceutical companies
CalTech	California Institute of Technology
CERN	European laboratory for particle physics
CGIAR	Consultative Group for International Agriculture Research
Chemistry	Nobel Prize for chemistry
CNRS	National Centre for Scientific Research, France
CSHL	Cold Spring Harbor Laboratory, Long Island, New York
CSIRO	Commonwealth Scientific and Industrial Research Organisation, Australia
CSL	Commonwealth Serum Laboratories
Economics	Bank of Sweden Prize in Economic Sciences in Memory of Alfred Nobel
EMBO	European Molecular Biology Organization
FAO	Food and Agriculture Organization of the United Nations
FDA	Food and Drug Administration, US

FRS	Fellow of the Royal Society of London
HHMI	Howard Hughes Medical Research Institute
ILRAD	International Laboratory for Research in Animal Diseases, Nairobi, Kenya
ILRI	International Livestock Research Institute, the successor to ILRAD
JCSMR	John Curtin School of Medical Research at the ANU, Canberra
LMB	Laboratory of Molecular Biology, Cambridge, England
MD	Doctor of Medicine
Medicine	Nobel Prize for Physiology or Medicine
MIT	Massachusetts Institute of Technology, Boston
MRC	British Medical Research Council
MSTP	Joint MD/PhD medical scientist training program, US
NAS	National Academy of Sciences, US
NCI	National Cancer Institute of the NIH, US
NHMRC	National Health and Medical Research Council, Australia
NIH	National Institutes of Health of the US Public Health Service
NSF	National Science Foundation, US
Peace	Nobel Peace Prize
Penn	University of Pennsylvania, Philadelphia
PhD	Doctor of Philosophy, the normal 'ticket' for research scientists
Physics	Nobel Prize for Physics
PI	Principal Investigator, the leader of a research program

Postdoc	Postdoctoral Fellow, normally a trainee position that follows the PhD
R&D	Research and development
RS	Royal Society of London, the British national academy of science
UCLA	University of California, Los Angeles
WEHI	Walter and Eliza Hall Institute for Medical Research, Melbourne
WHO	World Health Organization

Recommended Reading

The following provide useful insights into how science works.

Barry, John M, *The Great Influenza: The Epic Story of the Deadliest Plague in History*, Viking, New York, 2004.
The latest, and perhaps the best, account of the catastrophic 1918–19 influenza pandemic. In addition to describing the medical, social and political consequences of the massive debility and mortality (perhaps 40 million deaths) that occurred, the author also traces the history of US academic medicine. This is a fascinating, informative and readable book.

Gleick, James M, *Genius: Richard Feynman and Modern Physics*, Little Brown, London 1992.
This is the very readable, personal story of the extraordinary theoretical physicist who shared the 1965 Nobel Prize for Physics. A colourful, highly interactive character who was also a great teacher, Feynman led the enquiry into the *Challenger* shuttle disaster and cut through much obfuscation and hypocrisy to find the cause of the catastrophe.

Koestler, Arthur, *The Act of Creation*, Hutchison, London, 1976.
The first discussion of the creative process that many of my generation encountered, this book is still worth reading.

Kuhn, Thomas, *The Structure of Scientific Revolutions*, University of Chicago Press, Chicago, 1996.
Kuhn emphasised that science progresses by revolutions in understanding, which he called 'paradigm shifts'. Being a practitioner rather than a philosopher of science, I've found the viewpoints of both Kuhn and Popper (see below) to be useful, though some philosophers seem to consider that they are at odds.

Maddox, Brenda M, *Rosalind Franklin: The Dark Lady of DNA*, Harper Collins, London, 2002.
The X-ray crystallographer Rosalind Franklin generated the photographic image that was seen by Watson and Crick, without her knowledge, and allowed them to clarify their thinking about the nature of DNA. If she had not died young (of cancer) before their 1961 Nobel Medicine award, there would have been a problem with the Nobel 'rule of three', as her senior colleague, Maurice Wilkins, was included. Rosalind Franklin suffered from the blatant discrimination against women that was still problematic in some British academic institutions of that time.

The Nobel website: http://nobelprize.org/
The Nobel e-museum provides a comprehensive source of information about these awards.

Popper, Karl R, *Conjectures and Refutations: The Growth of Scientific Knowledge*, Routledge, London, 1989.
Popper transposed R.A. Fisher's statistical ideas to argue that experimental science progresses not so much by proving that A (the test group) is definitely different from B (the control), but rather by achieving the much easier goal of falsifying the null hypothesis that A and B are the same. This may sound a bit dry, but it is actually quite readable and a

good introduction to the mechanisms that scientists use to stay honest and avoid deluding themselves. A scientist who sets out to prove a point no matter what the cost can fall into grievous error.

Sobel, Dava, *Galileo's Daughter: A Drama of Science, Faith and Love*, Fourth Estate, London, 1999.
The author uses previously untranslated letters from Galileo's illegitimate daughter, Virginia, to develop the thesis that Galileo's condemnation and house arrest for arguing the Copernican view—that the Earth orbits the sun—had as much to do with ecclesiastical politics as with a desire to suppress the truth. Nonetheless, this move to suppress the truth did immense damage both to the church and to the renaissance in the Catholic countries of Europe.

Sobel, Dava, *Longitude: The Story of a Lone Genius who Solved the Greatest Scientific Problem of his Time*, Fourth Estate, London, 1995.
John Harrison developed the first accurate chronometers to facilitate navigation, but due recognition was delayed because of his lack of 'establishment' credentials.

Watson, James D, *The Double Helix*, Penguin, London, 1998.
Watson's 1968 account of the events that led him and Frances Crick to build (in 1953) the DNA model that began the era of molecular medicine. Evidently Crick and others at the Laboratory for Molecular Biology in Cambridge did not like this book too much, but it does convey a real sense of the way the type of science that I'm familiar with is done. It's quite short, fun to read, though it was clearly unfair to Rosalind Franklin (see above); maybe the callow Watson was just a bit scared of the sophisticated Rosalind!

Index

Index

Index

(HIV), 49–50, 102, 223, 226; human papillomavirus, 168; lymphocytic choriomeningitis virus (LMCV), 121–2, 125; poliomyelitis, 97–8, 99, 131; smallpox, 97, 231; *see also* AIDS, influenza

Walter and Eliza Hall Institute (Melbourne), 163
Wambugu, Florence, 44–5, 46
Watson, James, 71, 173, 220, 236; *The Double Helix*, 246
Webby, Richard, 69
Webster, Rob, 69
Wellcome Trust, the, 180
Wettenhall, Dick, 77
White, Patrick, 156, 157
Whitworth, Judy, 164
Wigzell, Hans, 14, 36–7
Wilberforce, Samuel, 200
Wilczec, Frank, 58

Wiley, Don, 21, 127, 130
Wilkins, Maurice, 220, 246
Williams, Jody, 155, 243
Williams, Robin, xix–xx
Williams, Robyn, 35
Wilson, Ian, 130
Wooldridge, Michael, 153
Wright, Sewall, 120

Yalow, Rosalyn, 136
Yuan Tseh Lee, 161
Yunus, Muhammad, 144

Zernik, Fritz, 87
Zewail, Ahmed, 92
Zinkernagel, Katherin, 126
Zinkernagel, Rolf, 7, 20, 72, 85, 105, 122–5, 164, 235, 246–7; and Nobel Prize, 1, 11, 86, 94, 113, 121, 126, 130, 215
Zur Hausen, Harald, 168
Zymborska, Wislawa, 13